Graduate Texts in Mathematics 9

Managing Editor: P. R. Halmos

J. E. Humphreys

Introduction to
Lie Algebras and
Representation Theory

Springer-Verlag New York · Heidelberg · Berlin

1972

J. E. Humphreys

Associate Professor of Mathematics, Courant Institute of Mathematical Sciences,
New York University

AMS Subject Classification (1970)
17 B XX, 20 G 05, 22 E 60

ISBN 0-387-90052-7 Springer-Verlag New York · Heidelberg · Berlin (soft
cover)
ISBN 0-387-90053-5 Springer-Verlag New York · Heidelberg · Berlin (hard
cover)
ISBN 3-540-90052-7 Springer-Verlag Berlin · Heidelberg · New York (soft
cover)

To the memory of my nephews

Willard Charles Humphreys III

and

Thomas Edward Humphreys

Preface

This book is designed to introduce the reader to the theory of semisimple Lie algebras over an algebraically closed field of characteristic 0, with emphasis on representations. A good knowledge of linear algebra (including eigenvalues, bilinear forms, euclidean spaces, and tensor products of vector spaces) is presupposed, as well as some acquaintance with the methods of abstract algebra. The first four chapters might well be read by a bright undergraduate; however, the remaining three chapters are admittedly a little more demanding.

Besides being useful in many parts of mathematics and physics, the theory of semisimple Lie algebras is inherently attractive, combining as it does a certain amount of depth and a satisfying degree of completeness in its basic results. Since Jacobson's book appeared a decade ago, improvements have been made even in the classical parts of the theory. I have tried to incorporate some of them here and to provide easier access to the subject for non-specialists. For the specialist, the following features should be noted:

(1) The Jordan-Chevalley decomposition of linear transformations is emphasized, with "toral" subalgebras replacing the more traditional Cartan subalgebras in the semisimple case.

(2) The conjugacy theorem for Cartan subalgebras is proved (following D. J. Winter and G. D. Mostow) by elementary Lie algebra methods, avoiding the use of algebraic geometry.

(3) The isomorphism theorem is proved first in an elementary way (Theorem 14.2), but later obtained again as a corollary of Serre's Theorem (18.3), which gives a presentation by generators and relations.

(4) From the outset, the simple algebras of types A, B, C, D are emphasized in the text and exercises.

(5) Root systems are treated axiomatically (Chapter III), along with some of the theory of weights.

(6) A conceptual approach to Weyl's character formula, based on Harish-Chandra's theory of "characters" and independent of Freudenthal's multiplicity formula (22.3), is presented in §23 and §24. This is inspired by D.-N. Verma's thesis, and recent work of I. N. Bernstein, I. M. Gel'fand, S. I. Gel'fand.

(7) The basic constructions in the theory of Chevalley groups are given in Chapter VII, following lecture notes of R. Steinberg.

I have had to omit many standard topics (most of which I feel are better suited to a second course), e.g., cohomology, theorems of Levi and Mal'cev, theorems of Ado and Iwasawa, classification over non-algebraically closed fields, Lie algebras in prime characteristic. I hope the reader will be stimulated to pursue these topics in the books and articles listed under References, especially Jacobson [1], Bourbaki [1], [2], Winter [1], Seligman [1].

A few words about mechanics: Terminology is mostly traditional, and notation has been kept to a minimum, to facilitate skipping back and forth in the text. After Chapters I–III, the remaining chapters can be read in almost any order if the reader is willing to follow up a few references (except that VII depends on §20 and §21, while VI depends on §17). A reference to Theorem 14.2 indicates the (unique) theorem in subsection 14.2 (of §14). Notes following some sections indicate nonstandard sources or further reading, but I have not tried to give a history of each theorem (for historical remarks, cf. Bourbaki [2] and Freudenthal-deVries [1]). The reference list consists largely of items mentioned explicitly; for more extensive bibliographies, consult Jacobson [1], Seligman [1]. Some 240 exercises, of all shades of difficulty, have been included; a few of the easier ones are needed in the text.

This text grew out of lectures which I gave at the N.S.F. Advanced Science Seminar on Algebraic Groups at Bowdoin College in 1968; my intention then was to enlarge on J.-P. Serre's excellent but incomplete lecture notes [2]. My other literary debts (to the books and lecture notes of N. Bourbaki, N. Jacobson, R. Steinberg, D. J. Winter, and others) will be obvious. Less obvious is my personal debt to my teachers, George Seligman and Nathan Jacobson, who first aroused my interest in Lie algebras. I am grateful to David J. Winter for giving me pre-publication access to his book, to Robert L. Wilson for making many helpful criticisms of an earlier version of the manuscript, to Connie Engle for her help in preparing the final manuscript, and to Michael J. DeRise for moral support. Financial assistance from the Courant Institute of Mathematical Sciences and the National Science Foundation is also gratefully acknowledged.

New York, April 4, 1972   J. E. Humphreys

Notation and Conventions

$\mathbf{Z}$, $\mathbf{Z}^+$, $\mathbf{Q}$, $\mathbf{R}$, $\mathbf{C}$ denote (respectively) the integers, nonnegative integers, rationals, reals, and complex numbers

$\amalg$ denotes direct sum of vector spaces

$A \ltimes B$ denotes the semidirect product of groups A and B, with B normal

Card = cardinality

char = characteristic

det = determinant

dim = dimension

Ker = kernel

Im   = image

Tr   = trace

Table of Contents

Introduction to Lie Algebras and Representation Theory

Chapter I

Basic Concepts

In this chapter F *denotes an arbitrary (commutative) field.*

1. Definitions and first examples

1.1. The notion of Lie algebra

Lie algebras arise "in nature" as vector spaces of linear transformations endowed with a new operation which is in general neither commutative nor associative: $[x, y] = xy - yx$ (where the operations on the right side are the usual ones). It is possible to describe this kind of system abstractly in a few axioms.

Definition. A vector space L over a field F, with an operation $L \times L \to L$, denoted $(x, y) \mapsto [xy]$ and called the **bracket** or **commutator** of x and y, is called a **Lie algebra** over F if the following axioms are satisfied:

(L1) The bracket operation is bilinear.

(L2) $[xx] = 0$ for all x in L.

(L3) $[x[yz]] + [y[zx]] + [z[xy]] = 0 \quad (x, y, z \in L)$.

$$[[yz], x] = [[xz], y] + [[yx], z]$$
$$[x[yz]] = [y[xz]] + [z[yx]]$$

Axiom $(L3)$ is called the **Jacobi identity.** Notice that $(L1)$ and $(L2)$, applied to $[x+y, x+y]$, imply anticommutativity: $(L2')$ $[xy] = -[yx]$. (Conversely, if char F $\neq 2$, it is clear that $(L2')$ will imply $(L2)$.)

We say that two Lie algebras L, L' over F are **isomorphic** if there exists a vector space isomorphism $\phi: L \to L'$ satisfying $\phi([xy]) = [\phi(x)\phi(y)]$ for all x, y in L (and then ϕ is called an **isomorphism** of Lie algebras). Similarly, it is obvious how to define the notion of (Lie) **subalgebra** of L: A subspace K of L is called a subalgebra if $[xy] \in K$ whenever $x, y \in K$; in particular, K is a Lie algebra in its own right relative to the inherited operations. Note that any nonzero element $x \in L$ defines a one dimensional subalgebra Fx, with trivial multiplication, because of $(L2)$.

In this book we shall be concerned almost exclusively with Lie algebras L whose underlying vector space is *finite dimensional* over F. *This will always be assumed, unless otherwise stated.* We hasten to point out, however, that certain infinite dimensional vector spaces and associative algebras over F will play a vital role in the study of representations (Chapters V–VII). We also mention, before looking at some concrete examples, that the axioms for a Lie algebra make perfectly good sense if L is only assumed to be a module over a commutative ring, but we shall not pursue this point of view here.

1

1.2. Linear Lie algebras

If V is a finite dimensional vector space over F, denote by End V the set of linear transformations $V \to V$. As a vector space over F, End V has dimension n^2 ($n = \dim V$), and End V is a ring relative to the usual product operation. Define a new operation $[x, y] = xy - yx$, called the **bracket** of x and y. With this operation End V becomes a Lie algebra over F: axioms (L1) and (L2) are immediate, while (L3) requires a brief calculation (which the reader is urged to carry out at this point). In order to distinguish this new algebra structure from the old associative one, we write $\mathfrak{gl}(V)$ for End V viewed as Lie algebra and call it the **general linear algebra** (because it is closely associated with the **general linear group** $GL(V)$ consisting of all invertible endomorphisms of V). When V is infinite dimensional, we shall also use the notation $\mathfrak{gl}(V)$ without further comment.

Any subalgebra of a Lie algebra $\mathfrak{gl}(V)$ is called a **linear Lie algebra**. The reader who finds matrices more congenial than linear transformations may prefer to fix a basis for V, thereby identifying $\mathfrak{gl}(V)$ with the set of all $n \times n$ matrices over F, denoted $\mathfrak{gl}(n, \text{F})$. This procedure is harmless, and very convenient for making explicit calculations. For reference, we write down the multiplication table for $\mathfrak{gl}(n, \text{F})$ relative to the standard basis consisting of the matrices e_{ij} (having 1 in the (i, j) position and 0 elsewhere). Since $e_{ij}e_{kl} = \delta_{jk}e_{il}$, it follows that:

(*) $$[e_{ij}, e_{kl}] = \delta_{jk}\,e_{il} - \delta_{li}\,e_{kj}.$$

Notice that the coefficients are all ± 1 or 0; in particular, all of them lie in the prime field of F.

Now for some further examples, which are central to the theory we are going to develop in this book. They fall into four families A_ℓ, B_ℓ, C_ℓ, D_ℓ ($\ell \geq 1$) and are called the **classical algebras** (because they correspond to certain of the classical linear Lie groups).

A_ℓ: Let $\dim V = \ell + 1$. Denote by $\mathfrak{sl}(V)$, or $\mathfrak{sl}(\ell+1, \text{F})$, the set of endomorphisms of V having trace zero. (Recall that the **trace** of a matrix is the sum of its diagonal entries; this is independent of choice of basis for V, hence makes sense for an endomorphism of V.) Since $Tr(xy) = Tr(yx)$, and $Tr(x+y) = Tr(x) + Tr(y)$, $\mathfrak{sl}(V)$ is a subalgebra of $\mathfrak{gl}(V)$, called the **special linear algebra** because of its connection with the **special linear group** $SL(V)$ of endomorphisms of det 1. What is its dimension? On the one hand $\mathfrak{sl}(V)$ is a proper subalgebra of $\mathfrak{gl}(V)$, hence of dimension at most $(\ell+1)^2 - 1$. On the other hand, we can exhibit this number of linearly independent matrices of trace zero: Take all e_{ij} ($i \neq j$), along with all $h_i = e_{ii} - e_{i+1,i+1}$ ($1 \leq i \leq \ell$), for a total of $\ell + (\ell+1)^2 - (\ell+1)$ matrices. We shall always view this as the standard basis for $\mathfrak{sl}(\ell+1, \text{F})$.

C_ℓ: Let $\dim V = 2\ell$, with basis $(v_1, \ldots, v_{2\ell})$. Define a nondegenerate skew-symmetric form f on V by the matrix $s = \begin{pmatrix} 0 & I_\ell \\ -I_\ell & 0 \end{pmatrix}$. (It can be shown

$$f(v, w) = v^{\mathsf{T}} s w$$

that even dimensionality is a necessary condition for existence of a non-degenerate bilinear form satisfying $f(v, w) = -f(w, v)$.) Denote by $\mathfrak{sp}(V)$, or $\mathfrak{sp}(2\ell, F)$, the **symplectic algebra**, which by definition consists of all endomorphisms x of V satisfying $f(x(v), w) = -f(v, x(w))$. The reader can easily verify that $\mathfrak{sp}(V)$ is closed under the bracket operation. In matrix terms, the condition for $x = \begin{pmatrix} m & n \\ p & q \end{pmatrix}$ $(m, n, p, q \in \mathfrak{gl}(\ell, F))$ to be symplectic is that $sx = -x^t s$ (x^t = transpose of x), i.e., that $n^t = n$, $p^t = p$, and $m^t = -q$. (This last condition forces $Tr(x) = 0$.) It is easy now to compute a basis for $\mathfrak{sp}(2\ell, F)$. Take the diagonal matrices $e_{ii} - e_{\ell+i,\ell+i}$ $(1 \le i \le \ell)$, ℓ in all. Add to these all $e_{ij} - e_{\ell+j,\ell+i}$ $(1 \le i \ne j \le \ell)$, $\ell^2 - \ell$ in number. For n we use the matrices $e_{i,\ell+i}$ $(1 \le i \le \ell)$ and $e_{i,\ell+j} + e_{j,\ell+i}$ $(1 \le i < j \le \ell)$, a total of $\ell + \frac{1}{2}\ell(\ell-1)$, and similarly for the positions in p. Adding up, we find dim $\mathfrak{sp}(2\ell, F) = 2\ell^2 + \ell$.

B_ℓ: Let dim $V = 2\ell+1$ be odd, and take f to be the nondegenerate symmetric bilinear form on V whose matrix is $s = \begin{pmatrix} 1 & 0 & 0 \\ 0 & 0 & I_\ell \\ 0 & I_\ell & 0 \end{pmatrix}$. The **orthogonal algebra** $\mathfrak{o}(V)$, or $\mathfrak{o}(2\ell+1, F)$, consists of all endomorphisms of V satisfying $f(x(v), w) = -f(v, x(w))$ (the same requirement as for C_ℓ). If we partition x in the same form as s, say $x = \begin{pmatrix} a & b_1 & b_2 \\ c_1 & m & n \\ c_2 & p & q \end{pmatrix}$, then the condition $sx = -x^t s$ translates into the following set of conditions: $a = 0$, $c_1 = -b_2^t$, $c_2 = -b_1^t$, $q = -m^t$, $n^t = -n$, $p^t = -p$. (As in the case of C_ℓ, this shows that $Tr(x) = 0$.) For a basis, take first the ℓ diagonal matrices $e_{ii} - e_{\ell+i,\ell+i}$ $(2 \le i \le \ell+1)$. Add the 2ℓ matrices involving only row one and column one: $e_{1,\ell+i+1} - e_{i+1,1}$ and $e_{1,i+1} - e_{\ell+i+1,1}$ $(1 \le i \le \ell)$. Corresponding to $q = -m^t$, take (as for C_ℓ) $e_{i+1,j+1} - e_{\ell+j+1,\ell+i+1}$ $(1 \le i \ne j \le \ell)$. For n take $e_{i+1,\ell+j+1} - e_{j+1,\ell+i+1}$ $(1 \le i < j \le \ell)$, and for p, $e_{i+\ell+1,j+1} - e_{j+\ell+1,i+1}$ $(1 \le j < i \le \ell)$. The total number of basis elements is $2\ell^2 + \ell$ (notice that this was also the dimension of C_ℓ).

D_ℓ: Here we obtain another **orthogonal algebra**. The construction is identical to that for B_ℓ, except that dim $V = 2\ell$ is even and s has the simpler form $\begin{pmatrix} 0 & I_\ell \\ I_\ell & 0 \end{pmatrix}$. We leave it as an exercise for the reader to construct a basis and to verify that dim $\mathfrak{o}(2\ell, F) = 2\ell^2 - \ell$ (Exercise 8).

We conclude this subsection by mentioning several other subalgebras of $\mathfrak{gl}(n, F)$ which play an important subsidiary role for us. Let $\mathfrak{t}(n, F)$ be the set of **upper triangular matrices** (a_{ij}), $a_{ij} = 0$ if $i > j$. Let $\mathfrak{n}(n, F)$ be the **strictly upper triangular matrices** (a_{ij}), $a_{ij} = 0$ if $i \ge j$. Finally, let $\mathfrak{d}(n, F)$ be the set of all **diagonal matrices**. It is trivial to check that each of these is closed under the bracket. Notice also that $\mathfrak{t}(n, F) = \mathfrak{d}(n, F) + \mathfrak{n}(n, F)$ (vector space direct sum), with $[\mathfrak{d}(n, F), \mathfrak{n}(n, F)] = \mathfrak{n}(n, F)$, hence $[\mathfrak{t}(n, F), \mathfrak{t}(n, F)] = \mathfrak{n}(n, F)$, cf. Exercise 5. (If H, K are subalgebras of L, $[H\,K]$ denotes the subspace of L spanned by commutators $[xy]$, $x \in H$, $y \in K$.)

1.3. Lie algebras of derivations

Some Lie algebras of linear transformations arise most naturally as derivations of algebras. By an **F-algebra** (not necessarily associative) we simply mean a vector space $\mathfrak{A}$ over F endowed with a bilinear operation $\mathfrak{A} \times \mathfrak{A} \to \mathfrak{A}$, usually denoted by juxtaposition (unless $\mathfrak{A}$ is a Lie algebra, in which case we always use the bracket). By a **derivation** of $\mathfrak{A}$ we mean a linear map $\delta \colon \mathfrak{A} \to \mathfrak{A}$ satisfying the familiar product rule $\delta(ab) = a\delta(b) + \delta(a)b$. It is easily checked that the collection Der $\mathfrak{A}$ of all derivations of $\mathfrak{A}$ is a vector subspace of End $\mathfrak{A}$. The reader should also verify that the commutator $[\delta, \delta']$ of two derivations is again a derivation (though the ordinary product need not be, cf. Exercise 11). So Der $\mathfrak{A}$ is a subalgebra of $\mathfrak{gl}(\mathfrak{A})$.

Since a Lie algebra L is an F-algebra in the above sense, Der L is defined. Certain derivations arise quite naturally, as follows. If $x \in L$, $y \mapsto [xy]$ is an endomorphism of L, which we denote ad x. In fact, ad $x \in$ Der L, because we can rewrite the Jacobi identity (using $(L2')$) in the form: $[x[yz]] = [[xy]z] + [y[xz]]$. Derivations of this form are called **inner**, all others **outer**. It is of course perfectly possible to have ad $x = 0$ even when $x \neq 0$: this occurs in any one dimensional Lie algebra, for example. The map $L \to$ Der L sending x to ad x is called the **adjoint representation** of L; it plays a decisive role in all that follows.

Sometimes we have occasion to view x simultaneously as an element of L and of a subalgebra K of L. To avoid ambiguity, the notation $\text{ad}_L x$ or $\text{ad}_K x$ will be used to indicate that x is acting on L (respectively, K). For example, if x is a diagonal matrix, then $\text{ad}_{\mathfrak{d}(n,\mathsf{F})}(x) = 0$, whereas $\text{ad}_{\mathfrak{gl}(n,\mathsf{F})}(x)$ need not be zero.

1.4. Abstract Lie algebras

We have looked at some natural examples of linear Lie algebras. It is known that, in fact, every (finite dimensional) Lie algebra is isomorphic to some linear Lie algebra (theorems of Ado, Iwasawa). This will not be proved here (cf. Jacobson [1] Chapter VI, or Bourbaki [1]); however, it will be obvious at an early stage of the theory that the result is true for all cases we are interested in.

Sometimes it is desirable, however, to contemplate Lie algebras abstractly. For example, if L is an arbitrary finite dimensional vector space over F, we can view L as a Lie algebra by setting $[xy] = 0$ for all x, $y \in L$. Such an algebra, having trivial Lie multiplication, is called **abelian** (because in the linear case $[x, y] = 0$ just means that x and y commute). If L is any Lie algebra, with basis $x_1, \ldots, x_n$ it is clear that the entire multiplication table of L can be recovered from the **structure constants** a_{ij}^k which occur in the expressions $[x_i x_j] = \sum_{k=1}^{n} a_{ij}^k x_k$. Those for which $i \geq j$ can even be deduced from the others, thanks to $(L2)$, $(L2')$. Turning this remark around, it is possible to define an abstract Lie algebra from scratch simply by specifying a set of structure constants. Naturally, not just any set of scalars $\{a_{ij}^k\}$ will do, but a moment's thought shows that it is enough to require the "obvious" identities, those implied by (L2) and (L3):

$$a_{ii}^k = 0;$$

$$\sum_k (a_{ij}^k a_{k\ell}^m + a_{j\ell}^k a_{ki}^m + a_{\ell i}^k a_{kj}^m) = 0.$$

In practice, we shall have no occasion to construct Lie algebras in this artificial way. But, as an application of the abstract point of view, we can determine (up to isomorphism) all Lie algebras of dimension ≤ 2. In dimension 1 there is a single basis vector x, with multiplication table $[xx] = 0$ (L2). In dimension 2, start with a basis x, y of L. Clearly, all products in L yield scalar multiples of $[xy]$. If these are all 0, then L is abelian. Otherwise, we can replace x in the basis by a vector spanning the one dimensional space of multiples of the original $[xy]$, and take y to be any other vector independent of the new x. Then $[xy] = ax$ ($a \neq 0$). Replacing y by $a^{-1}y$, we finally get $[xy] = x$. Abstractly, therefore, at most one nonabelian L exists (the reader should check that $[xy] = x$ actually defines a Lie algebra).

Exercises

1. Let L be the real vector space $\mathbf{R}^3$. Define $[xy] = x \times y$ (cross product of vectors) for $x, y \in L$, and verify that L is a Lie algebra. Write down the structure constants relative to the usual basis of $\mathbf{R}^3$.

2. Verify that the following equations and those implied by (L1) (L2) define a Lie algebra structure on a three dimensional vector space with basis (x, y, z): $[xy] = z$, $[xz] = y$, $[yz] = 0$.

3. Let $x = \begin{pmatrix} 0 & 1 \\ 0 & 0 \end{pmatrix}$, $h = \begin{pmatrix} 1 & 0 \\ 0 & -1 \end{pmatrix}$, $y = \begin{pmatrix} 0 & 0 \\ 1 & 0 \end{pmatrix}$ be an ordered basis for $\mathfrak{sl}(2, \mathsf{F})$. Compute the matrices of ad x, ad h, ad y relative to this basis.

4. Find a linear Lie algebra isomorphic to the nonabelian two dimensional algebra constructed in (1.4). [Hint: Look at the adjoint representation.]

5. Verify the assertions made in (1.2) about $\mathfrak{t}(n, \mathsf{F})$, $\mathfrak{d}(n, \mathsf{F})$, $\mathfrak{n}(n, \mathsf{F})$, and compute the dimension of each algebra, by exhibiting bases.

6. Let $x \in \mathfrak{gl}(n, \mathsf{F})$ have n *distinct* eigenvalues $a_1, \ldots, a_n$ in F. Prove that the eigenvalues of ad x are precisely the n^2 scalars $a_i - a_j$ ($1 \leq i, j \leq n$), which of course need not be distinct.

7. Let $\mathfrak{s}(n, \mathsf{F})$ denote the **scalar matrices** ($=$ scalar multiples of the identity) in $\mathfrak{gl}(n, \mathsf{F})$. If char F is 0 or else a prime not dividing n, prove that $\mathfrak{gl}(n, \mathsf{F}) = \mathfrak{sl}(n, \mathsf{F}) + \mathfrak{s}(n, \mathsf{F})$ (direct sum of vector spaces), with $[\mathfrak{s}(n, \mathsf{F}), \mathfrak{gl}(n, \mathsf{F})] = 0$.

8. Verify the stated dimension of D_ℓ.

9. When char $\mathsf{F} = 0$, show that each classical algebra $L = \mathsf{A}_\ell$, B_ℓ, C_ℓ, or D_ℓ is equal to $[LL]$. (This shows again that each algebra consists of trace 0 matrices.)

10. For small values of ℓ, isomorphisms occur among certain of the classical algebras. Show that A_1, B_1, C_1 are all isomorphic, while D_1 is the one dimensional Lie algebra. Show that B_2 is isomorphic to C_2, D_3 to A_3. What can you say about D_2?

11. Verify that the commutator of two derivations of an F-algebra is again a derivation, whereas the ordinary product need not be.

12. For each classical linear Lie algebra $L \subset \mathfrak{gl}(V)$, as described in (1.2), verify that relative to any choice of basis for V, transposes of the matrices of L all lie in L. [This is obvious for A_ℓ. For C_ℓ, notice that $-s = s^{-1}$, while for B_ℓ and D_ℓ, $s = s^{-1}$.]

2. Ideals and homomorphisms

2.1. Ideals

A subspace I of a Lie algebra L is called an **ideal** of L if $x \in L$, $y \in I$ together imply $[xy] \in I$. (Since $[xy] = -[yx]$, the condition could just as well be written: $[yx] \in I$.) Ideals play the role in Lie algebra theory which is played by normal subgroups in group theory and by two sided ideals in ring theory: they arise as kernels of homomorphisms (2.2).

Obviously 0 (the subspace consisting only of the zero vector) and L itself are ideals of L. A less trivial example is the **center** $Z(L) = \{z \in L | [xz] = 0 \text{ for all } x \in L\}$. The Jacobi identity shows at once that $Z(L)$ is actually an ideal. Notice that L is abelian if and only if $Z(L) = L$. Another important example is the **derived algebra** of L, denoted $[LL]$, which is analogous to the commutator subgroup of a group. It consists of all linear combinations of commutators $[xy]$, and is clearly an ideal.

Evidently L is abelian if and only if $[LL] = 0$. At the other extreme, a study of the multiplication table for $L = \mathfrak{sl}(n, \text{F})$ in (1.2) ($n \neq 2$ if char $\text{F} = 2$) shows that $L = [LL]$ in this case, and similarly for other classical linear Lie algebras (Exercise 1.9).

If I, J are two ideals of a Lie algebra L, then $I + J = \{x + y | x \in I, y \in J\}$ is also an ideal. Similarly, $[IJ] = \{\sum x_i y_i | x_i \in I, y_i \in J\}$ is an ideal; the derived algebra $[LL]$ is just a special case of this construction.

It is natural to analyze the structure of a Lie algebra by looking at its ideals. If L has no ideals except itself and 0, and if moreover $[LL] \neq 0$, we call L **simple**. The condition $[LL] \neq 0$ (i.e., L nonabelian) is imposed in order to avoid giving undue prominence to the one dimensional algebra. Clearly, L simple implies $Z(L) = 0$ and $L = [LL]$.

Example. Let $L = \mathfrak{sl}(2, \text{F})$, char $\text{F} \neq 2$. Take as standard basis for L the three matrices (cf. (1.2)): $x = \begin{pmatrix} 0 & 1 \\ 0 & 0 \end{pmatrix}$, $y = \begin{pmatrix} 0 & 0 \\ 1 & 0 \end{pmatrix}$, $h = \begin{pmatrix} 1 & 0 \\ 0 & -1 \end{pmatrix}$. The multiplication table is then completely determined by the equations: $[xy] = h$, $[hx] = 2x$, $[hy] = -2y$. (Notice that x, y, h are eigenvectors for ad h, corresponding to the eigenvalues $2, -2, 0$. Since char $\text{F} \neq 2$, these eigenvalues are distinct.) If $I \neq 0$ is an ideal of L, let $ax + by + ch$ be an arbitrary nonzero element of I. Applying ad x twice, we get $-2bx \in I$, and applying ad y twice, we get $-2ay \in I$. Therefore, if a or b is nonzero, I contains either y or x (char $\mathsf{F} \neq 2$), and then, clearly, $I = L$ follows. On the other hand, if $a = b = 0$, then $0 \neq ch \in I$, so $h \in I$, and again $I = L$ follows. We conclude that *L is simple*.

In case a Lie algebra L is not simple (and not one dimensional) it is possible to "factor out" a nonzero proper ideal I and thereby obtain a Lie algebra of smaller dimension. The construction of a **quotient algebra** L/I (I an ideal of L) is formally the same as the construction of a quotient ring: as vector space L/I is just the quotient space, while its Lie multiplication is defined by $[x+I, y+I] = [xy]+I$. This is unambiguous, since if $x+I = x'+I$, $y+I = y'+I$, then we have $x' = x+u$ ($u \in I$), $y' = y+v$ ($v \in I$), whence $[x'y'] = [xy]+([uy]+[xv]+[uv])$, and therefore $[x'y']+I = [xy]+I$, since the terms in parentheses all lie in I.

For later use we mention a couple of related notions, analogous to those which arise in group theory. The **normalizer** of a subalgebra (or just subspace) K of L is defined by $N_L(K) = \{x \in L | [xK] \subset K\}$. By the Jacobi identity, $N_L(K)$ is a subalgebra of L; it may be described verbally as the largest sub-algebra of L which includes K as an ideal (in case K is a subalgebra to begin with). If $K = N_L(K)$, we call K **self-normalizing**; some important examples of this behavior will emerge later. The **centralizer** of a subset X of L is $C_L(X) = \{x \in L | [xX] = 0\}$. Again by the Jacobi identity, $C_L(X)$ is a subalgebra of L. For example, $C_L(L) = Z(L)$.

2.2. Homomorphisms and representations

The definition should come as no surprise. A linear transformation $\phi: L \to L'$ (L, L' Lie algebras over F) is called a **homomorphism** if $\phi([xy]) = [\phi(x)\phi(y)]$, for all x, $y \in L$. ϕ is called a **monomorphism** if Ker $\phi = 0$, an **epimorphism** if Im $\phi = L'$, an **isomorphism** (as in (1.1)) if it is both mono- and epi-. The first interesting observation to make is that *Ker ϕ is an ideal of L*: indeed, if $\phi(x) = 0$, and if $y \in L$ is arbitrary, then $\phi([xy]) = [\phi(x)\phi(y)] = 0$. It is also apparent that Im ϕ is a subalgebra of L'. As in other algebraic theories, there is a natural 1–1 correspondence between homomorphisms and ideals: to ϕ is associated Ker ϕ, and to an ideal I is associated the **canonical map** $x \mapsto x+I$ of L onto L/I. It is left as an easy exercise for the reader to verify the standard homomorphism theorems:

Proposition. (a) *If $\phi: L \to L'$ is a homomorphism of Lie algebras, then $L/\mathrm{Ker}\ \phi \cong \mathrm{Im}\ \phi$. If I is any ideal of L included in $\mathrm{Ker}\ \phi$, there exists a unique homomorphism $\psi: L/I \to L'$ making the following diagram commute ($\pi = $ canonical map):*

(b) If I and J are ideals of L such that $I \subset J$, then J/I is an ideal of L/I and $(L/I)/(J/I)$ is naturally isomorphic to L/J.

(c) If I, J are ideals of L, there is a natural isomorphism between $(I+J)/J$ and $I/(I \cap J)$. ☐

A **representation** of a Lie algebra L is a homomorphism $\phi: L \to \mathfrak{gl}(V)$ (V = vector space over F). Although we require L to be finite dimensional, it is useful to allow V to be of arbitrary dimension: $\mathfrak{gl}(V)$ makes sense in any case. However, for the time being the only important example to keep in mind is the adjoint representation ad: $L \to \mathfrak{gl}(L)$ introduced in (1.3), which sends x to ad x, where ad $x(y) = [xy]$. (The image of ad is in Der $L \subset \mathfrak{gl}(L)$, but this does not concern us at the moment.) It is clear that ad is a linear transformation. To see that it preserves the bracket, we calculate:

$$
\begin{aligned}
[\text{ad } x, \text{ad } y] (z) &= \text{ad } x \text{ ad } y(z) - \text{ad } y \text{ ad } x(z) \\
&= \text{ad } x([yz]) - \text{ad } y([xz]) \\
&= [x[yz]] - [y[xz]] \\
&= [x[yz]] + [[xz]y] \qquad\qquad (L2') \\
&= [[xy]z] \qquad\qquad\qquad\qquad (L3) \\
&= \text{ad } [xy] (z).
\end{aligned}
$$

What is the kernel of ad? It consists of all $x \in L$ for which ad $x = 0$, i.e., for which $[xy] = 0$ (all $y \in L$). So Ker ad $= Z(L)$. This already has an interesting consequence: If L is simple, then $Z(L) = 0$, so that ad: $L \to \mathfrak{gl}(L)$ is a monomorphism. This means that *any simple Lie algebra is isomorphic to a linear Lie algebra*.

2.3. Automorphisms

An **automorphism** of L is an isomorphism of L onto itself. Aut L denotes the group of all such. Important examples occur when L is a linear Lie algebra $\subset \mathfrak{gl}(V)$. If $g \in GL(V)$ is any invertible endomorphism of V, and if moreover $gLg^{-1} = L$, then it is immediate that the map $x \mapsto gxg^{-1}$ is an automorphism of L. For instance, if $L = \mathfrak{gl}(V)$ or even $\mathfrak{sl}(V)$, the second condition is automatic, so we obtain in this way a large collection of automorphisms. (Cf. Exercise 12.)

Now specialize to the case: char F $= 0$. Suppose $x \in L$ is an element for which ad x is **nilpotent**, i.e., $(\text{ad } x)^k = 0$ for some $k > 0$. Then the usual exponential power series for a linear transformation over **C** makes sense over F, because it has only finitely many terms: exp (ad x) $= 1 + \text{ad } x + (\text{ad } x)^2/2!$ $+ (\text{ad } x)^3/3! + \ldots + (\text{ad } x)^{k-1}/(k-1)!$. We claim that exp (ad x) $\in$ Aut L. More generally, this is true if ad x is replaced by an arbitrary nilpotent derivation δ of L. For this, use the familiar *Leibniz rule*:

$$
\frac{\delta^n}{n!} (xy) = \sum_{i=0}^{n} (1/i!) (\delta^i x) (1/(n-i)!) (\delta^{n-i} y).
$$

Then calculate as follows: (say $\delta^k = 0$)

$$\exp \delta(x) \exp \delta(y) = \left(\sum_{i=0}^{k-1} \left(\frac{\delta^i x}{i!}\right)\right)\left(\sum_{j=0}^{k-1} \left(\frac{\delta^j y}{j!}\right)\right)$$

$$= \sum_{n=0}^{2k-2} \left(\sum_{i=0}^{n} \left(\frac{\delta^i x}{i!}\right)\left(\frac{\delta^{n-i} y}{(n-i)!}\right)\right)$$

$$= \sum_{n=0}^{2k-2} \frac{\delta^n(xy)}{n!} \qquad \text{(Leibniz)}$$

$$= \sum_{n=0}^{k-1} \frac{\delta^n(xy)}{n!} \qquad (\delta^k = 0)$$

$$= \exp \delta(xy).$$

The fact that $\exp \delta$ is invertible follows (in the usual way) by exhibiting the explicit inverse $1 - \eta + \eta^2 - \eta^3 + \ldots \pm \eta^{k-1}$, $\exp \delta = 1 + \eta$.

An automorphism of the form $\exp (\text{ad } x)$, ad x nilpotent, is called **inner**; more generally, the subgroup of Aut L generated by these is denoted Int L and its elements called inner automorphisms. It is a *normal* subgroup: If $\phi \in \text{Aut } L$, $x \in L$, then $\phi(\text{ad } x)\phi^{-1} = \text{ad } \phi(x)$, whence $\phi \exp (\text{ad } x)\phi^{-1} = \exp (\text{ad } \phi(x))$.

For example, let $L = \mathfrak{sl}(2, \mathsf{F})$, with standard basis (x, y, h). Define $\sigma = \exp \text{ad } x \cdot \exp \text{ad } (-y) \cdot \exp \text{ad } x$ (so $\sigma \in \text{Int } L$). It is easy to compute the effect of σ on the basis (Exercise 10): $\sigma(x) = -y$, $\sigma(y) = -x$, $\sigma(h) = -h$. In particular, σ has order 2. Notice that $\exp x$, $\exp (-y)$ are well defined elements of $SL(2, \mathsf{F})$, the group of 2×2 matrices of det 1, conjugation by which leaves L invariant (as noted at the start of this subsection), so the product $s = (\exp x)(\exp -y)(\exp x)$ induces an automorphism $z \mapsto szs^{-1}$ of L. A quick calculation shows that $s = \begin{pmatrix} 0 & 1 \\ -1 & 0 \end{pmatrix}$ and that conjugating by s has precisely the same effect on L as applying σ.

The phenomenon just observed is not accidental: If $L \subset \mathfrak{gl}(V)$ is an arbitrary linear Lie algebra (char $\mathsf{F} = 0$), and $x \in L$ is nilpotent, then we claim that

(*) $\qquad (\exp x) y (\exp x)^{-1} = \exp \text{ad } x (y)$ for all $y \in L$.

To prove this, notice that ad $x = \lambda_x + \rho_{-x}$, where λ_x, ρ_x denote left and right multiplication by x in the ring End V (these commute, of course, and are nilpotent). Then the usual rules of exponentiation show that $\exp \text{ad } x = \exp (\lambda_x + \rho_{-x}) = \exp \lambda_x \cdot \exp \rho_{-x} = \lambda_{\exp x} \cdot \rho_{\exp (-x)}$, which implies (*).

Exercises

1. Prove that the set of all inner derivations ad x, $x \in L$, is an ideal of Der L.
2. Show that $\mathfrak{sl}(n, \mathsf{F})$ is precisely the derived algebra of $\mathfrak{gl}(n, \mathsf{F})$ (cf. Exercise 1.9).

3. Prove that the center of $\mathfrak{gl}(n, \mathsf{F})$ equals $\mathfrak{s}(n, \mathsf{F})$ (the scalar matrices). Prove that $\mathfrak{sl}(n, \mathsf{F})$ has center 0, unless char F divides n, in which case the center is $\mathfrak{s}(n, \mathsf{F})$.

4. Show that (up to isomorphism) there is a unique Lie algebra over F of dimension 3 whose derived algebra has dimension 1 and lies in $Z(L)$.

5. Suppose dim $L = 3$, $L = [LL]$. Prove that L must be simple. [Observe first that any homomorphic image of L also equals its derived algebra.] Recover the simplicity of $\mathfrak{sl}(2, \mathsf{F})$, char $\mathsf{F} \neq 2$.

6. Prove that $\mathfrak{sl}(3, \mathsf{F})$ is simple, unless char $\mathsf{F} = 3$ (cf. Exercise 3). [Use the standard basis h_1, h_2, e_{ij} $(i \neq j)$. If $I \neq 0$ is an ideal, then I is the direct sum of eigenspaces for ad h_1 or ad h_2; compare the eigenvalues of ad h_1, ad h_2 acting on the e_{ij}.]

7. Prove that $\mathfrak{t}(n, \mathsf{F})$ and $\mathfrak{d}(n, \mathsf{F})$ are self-normalizing subalgebras of $\mathfrak{gl}(n, \mathsf{F})$, whereas $\mathfrak{n}(n, \mathsf{F})$ has normalizer $\mathfrak{t}(n, \mathsf{F})$.

8. Prove that in each classical linear Lie algebra (1.2), the set of diagonal matrices is a self-normalizing subalgebra, when char $\mathsf{F} = 0$.

9. Prove Proposition 2.2.

10. Let σ be the automorphism of $\mathfrak{sl}(2, \mathsf{F})$ defined in (2.3). Verify that $\sigma(x) = -y$, $\sigma(y) = -x$, $\sigma(h) = -h$.

11. If $L = \mathfrak{sl}(n, \mathsf{F})$, $g \in GL(n, \mathsf{F})$, prove that the map of L to itself defined by $x \mapsto -gx^t g^{-1}$ (x^t = transpose of x) belongs to Aut L. When $n = 2$, g = identity matrix, prove that this automorphism is inner.

12. Let L be an orthogonal Lie algebra (type B_ℓ or D_ℓ). If g is an **orthogonal** matrix, in the sense that g is invertible and $g^t s g = s$, prove that $x \mapsto gxg^{-1}$ defines an automorphism of L.

3. Solvable and nilpotent Lie algebras

3.1. Solvability

It is natural to study a Lie algebra L via its ideals. In this section we exploit the formation of derived algebras. First, define a sequence of ideals of L (the **derived series**) by $L^{(0)} = L$, $L^{(1)} = [LL]$, $L^{(2)} = [L^{(1)}L^{(1)}], \ldots, L^{(i)} = [L^{(i-1)}L^{(i-1)}]$. Call L **solvable** if $L^{(n)} = 0$ for some n. For example, abelian implies solvable, whereas simple algebras are definitely nonsolvable.

An example which turns out to be rather general is the algebra $\mathfrak{t}(n, \mathsf{F})$ of upper triangular matrices, which was introduced in (1.2). The obvious basis for $\mathfrak{t}(n, \mathsf{F})$ consists of the matrix units e_{ij} for which $i \leq j$; the dimension is $1+2+\ldots+n = n(n+1)/2$. To show that $L = \mathfrak{t}(n, \mathsf{F})$ is solvable we compute explicitly its derived series, using the formula for commutators in (1.2). In particular, we have $[e_{ii}, e_{il}] = e_{il}$ for $i < l$, which shows that $\mathfrak{n}(n, \mathsf{F}) \subset [LL]$, where $\mathfrak{n}(n, \mathsf{F})$ is the subalgebra of upper triangular nilpotent matrices. Since $\mathfrak{t}(n, \mathsf{F}) = \mathfrak{d}(n, \mathsf{F}) + \mathfrak{n}(n, \mathsf{F})$, and since $\mathfrak{d}(n, \mathsf{F})$ is abelian, we conclude that $\mathfrak{n}(n, \mathsf{F})$ is equal to the derived algebra of L (cf. Exercise 1.5). Working next inside the algebra $\mathfrak{n}(n, \mathsf{F})$, we have a natural notion of "level" for e_{ij}, namely

$j-i$. In the formula for commutators, assume that $i < j$, $k < l$. Without losing any products we may also require $i \neq l$. Then $[e_{ij}, e_{kl}] = e_{il}$ (if $j = k$) or 0 (otherwise). In particular, each e_{il} is commutator of two matrices whose levels add up to that of e_{il}. We conclude that $L^{(2)}$ is spanned by those e_{ij} of level ≥ 2, $L^{(i)}$ by those of level $\geq 2^{i-1}$. Finally, it is clear that $L^{(i)} = 0$ whenever $2^{i-1} > n-1$.

Next we assemble a few simple observations about solvability.

Proposition. *Let L be a Lie algebra.*

(a) If L is solvable, then so are all subalgebras and homomorphic images of L.

(b) If I is a solvable ideal of L such that L/I is solvable, then L itself is solvable.

(c) If I, J are solvable ideals of L, then so is $I+J$.

Proof. (a) From the definition, if K is a subalgebra of L, then $K^{(i)} \subset L^{(i)}$. Similarly, if $\phi \colon L \to M$ is an epimorphism, an easy induction on i shows that $\phi(L^{(i)}) = M^{(i)}$.

(b) Say $(L/I)^{(n)} = 0$. Applying part (a) to the canonical homomorphism $\pi \colon L \to L/I$, we get $\pi(L^{(n)}) = 0$, or $L^{(n)} \subset I = \mathrm{Ker}\ \pi$. Now if $I^{(m)} = 0$, the obvious fact that $(L^{(i)})^{(j)} = L^{(i+j)}$ implies that $L^{(n+m)} = 0$ (apply proof of part (a) to the situation $L^{(n)} \subset I$).

(c) One of the standard homomorphism theorems (Proposition 2.2 (c)) yields an isomorphism between $(I+J)/J$ and $I/(I \cap J)$. As a homomorphic image of I, the right side is solvable, so $(I+J)/J$ is solvable. Then so is $I+J$, by part (b) applied to the pair $I+J$, J. □

As a first application, let L be an arbitrary Lie algebra and let S be a maximal solvable ideal (i.e., one included in no larger solvable ideal). If I is any other solvable ideal of L, then part (c) of the Proposition forces $S+I = S$ (by maximality), or $I \subset S$. This proves the existence of a unique maximal solvable ideal, called the **radical** of L and denoted Rad L. If $L \neq 0$ and Rad $L = 0$, L is called **semisimple**. For example, *a simple algebra is semisimple*: L has no ideals except itself and 0, and L is nonsolvable. Notice that if L is not solvable, i.e., $L \neq$ Rad L, then $L/$Rad L is semisimple (use part (b) of the proposition). The study of semisimple Lie algebras (char F $= 0$) will occupy most of this book. (But certain solvable subalgebras will also be needed along the way.)

3.2. Nilpotency

The definition of solvability imitates the corresponding notion in group theory, which goes back to Abel and Galois. By contrast, the notion of nilpotent group is more recent, and is modeled on the corresponding notion for Lie algebras. Define a sequence of ideals of L (the **descending central series**, also called the **lower central series**) by $L^0 = L$, $L^1 = [LL]\ (=L^{(1)})$, $L^2 = [LL^1], \ldots, L^i = [L\ L^{i-1}]$. L is called **nilpotent** if $L^n = 0$ for some n. For example, any abelian algebra is nilpotent. Clearly, $L^{(i)} \subset L^i$ for all i, so nilpotent algebras are solvable. The converse is false, however. Consider again $L = \mathfrak{t}(n, \mathsf{F})$. Our discussion in (3.1) showed that $L^{(1)} = L^1$ is $\mathfrak{n}(n, \mathsf{F})$, and also that $L^2 = [L\, L^1] = L^1$, so $L^i = L^1$ for all $i \geq 1$. On the other hand, it is easy to see that $M = \mathfrak{n}(n, \mathsf{F})$ is nilpotent: M^1 is spanned by those e_{ij} of level 2, M^2 by those of level 3, ..., M^i by those of level $i+1$.

Proposition. *Let L be a Lie algebra.*

(a) *If L is nilpotent, then so are all subalgebras and homomorphic images of L.*

(b) *If $L/Z(L)$ is nilpotent, then so is L.*

(c) *If L is nilpotent, then $Z(L) \neq 0$.*

Proof. (a) Imitate the proof of Proposition 3.1 (a).

(b) Say $L^n \subset Z(L)$, then $L^{n+1} = [LL^n] \subset [LZ(L)] = 0$.

(c) The last nonzero term of the descending central series is central.   □

The condition for L to be nilpotent can be rephrased as follows: For some n (depending only on L), $\operatorname{ad} x_1 \operatorname{ad} x_2 \ldots \operatorname{ad} x_n(y) = 0$ for all x_i, $y \in L$. In particular, $(\operatorname{ad} x)^n = 0$ for all $x \in L$. Now if L is any Lie algebra, and $x \in L$, we call x **ad-nilpotent** if $\operatorname{ad} x$ is a nilpotent endomorphism. Using this language, our conclusion can be stated: If L is nilpotent, then all elements of L are ad-nilpotent. It is a pleasant surprise to find that the converse is also true.

Theorem (Engel). *If all elements of L are ad-nilpotent, then L is nilpotent.*

The proof will be given in the next subsection. Using Engel's Theorem, it is easy to prove that $\mathfrak{n}(n, \mathsf{F})$ is nilpotent, without actually calculating the descending central series. We need only apply the following simple lemma.

Lemma. *Let $x \in \mathfrak{gl}(V)$ be a nilpotent endomorphism. Then $\operatorname{ad} x$ is also nilpotent.*

Proof. As in (2.3), we may associate to x two endomorphisms of End V, left and right translation: $\lambda_x(y) = xy$, $\rho_x(y) = yx$, which are nilpotent because x is. Moreover λ_x and ρ_x obviously commute. In any ring (here End (End V)) the sum or difference of two commuting nilpotents is again nilpotent (why?), so $\operatorname{ad} x = \lambda_x - \rho_x$ is nilpotent.   □

A word of warning: It is easy for a matrix to be ad-nilpotent in $\mathfrak{gl}(n, \mathsf{F})$ without being nilpotent. (The identity matrix is an example.) The reader should keep in mind two contrasting types of nilpotent linear Lie algebras: $\mathfrak{d}(n, \mathsf{F})$ and $\mathfrak{n}(n, \mathsf{F})$.

3.3. Proof of Engel's Theorem

Engel's Theorem (3.2) will be deduced from the following result, which is of interest in its own right. Recall that a single nilpotent linear transformation always has at least one eigenvector, corresponding to its unique eigenvalue 0. This is just the case dim $L = 1$ of the following theorem.

Theorem. *Let L be a subalgebra of $\mathfrak{gl}(V)$, V finite dimensional. If L consists of nilpotent endomorphisms and $V \neq 0$, then there exists nonzero $v \in V$ for which $L.v = 0$.*

Proof. Use induction on dim L, the case dim $L = 0$ (or dim $L = 1$) being obvious. Suppose $K \neq L$ is any subalgebra of L. According to Lemma 3.2, K acts (via ad) as a Lie algebra of nilpotent linear transformations on the vector space L, hence also on the vector space L/K. Because dim $K <$ dim L, the induction hypothesis guarantees existence of a vector $x + K \neq K$ in L/K killed by the image of K in $\mathfrak{gl}(L/K)$. This just means that $[yx] \in K$ for all $y \in K$, whereas $x \notin K$. In other words, K is properly included in $N_L(K)$ (the normalizer of K in L, see (2.1)).

Now take K to be a maximal proper subalgebra of L. The preceding argument forces $N_L(K) = L$, i.e., K is an *ideal* of L. If dim L/K were greater than one, then the inverse image in L of a one dimensional subalgebra of L/K (which always exists) would be a proper subalgebra properly containing K, which is absurd; therefore, K has codimension one. This allows us to write $L = K + \mathsf{F}z$ for any $z \in L - K$.

By induction, $W = \{v \in V \mid K.v = 0\}$ is nonzero. Since K is an ideal, W is stable under L: $x \in L$, $y \in K$, $w \in W$ imply $yx.w = xy.w - [x, y].w = 0$. Choose $z \in L - K$ as above, so the nilpotent endomorphism z (acting now on the subspace W) has an eigenvector, i.e., there exists nonzero $v \in W$ for which $z.v = 0$. Finally, $L.v = 0$, as desired. ☐

Proof of Engel's Theorem. We are given a Lie algebra L all of whose elements are ad-nilpotent; therefore, the algebra ad $L \subset \mathfrak{gl}(L)$ satisfies the hypothesis of Theorem 3.1. (We can assume $L \neq 0$.) Conclusion: There exists $x \neq 0$ in L for which $[Lx] = 0$, i.e., $Z(L) \neq 0$. Now $L/Z(L)$ evidently consists of ad-nilpotent elements and has smaller dimension than L. Using induction on dim L, we find that $L/Z(L)$ is nilpotent. Part (b) of Proposition 3.2 then implies that L itself is nilpotent. ☐

There is a useful corollary (actually, an equivalent version) of Theorem 3.3, which shows how "typical" $\mathfrak{n}(n, \mathsf{F})$ is. First a definition: If V is a finite dimensional vector space (say dim $V = n$), a **flag** in V is a chain of subspaces $0 = V_0 \subset V_1 \subset \ldots \subset V_n = V$, dim $V_i = i$. If $x \in$ End V, we say x stabilizes (or leaves invariant) this flag provided $x.V_i \subset V_i$ for all i.

Corollary. *Under the hypotheses of the theorem there exists a flag (V_i) in V stable under L, with $x.V_i \subset V_{i-1}$ for all i. In other words, there exists a basis of V relative to which the matrices of L are all in $\mathfrak{n}(n, \mathsf{F})$.*

Proof. Begin with any nonzero $v \in V$ killed by L, the existence of which is assured by the theorem . Set $V_1 = \mathsf{F}v$. Let $W = V/V_1$, and observe that the induced action of L on W is also by nilpotent endomorphisms. By induction on dim V, W has a flag stabilized by L, whose inverse image in V does the trick. ☐

To conclude this section, we mention a typical application of Theorem 3.3, which will be needed later on.

Lemma. *Let L be nilpotent, K an ideal of L. Then if $K \neq 0$, $K \cap Z(L) \neq 0$. (In particular, $Z(L) \neq 0$; cf. Proposition 3.2(c).)*

Proof. L acts on K via the adjoint representation, so Theorem 3.3 yields nonzero $x \in K$ killed by L, i.e., $[Lx] = 0$, so $x \in K \cap Z(L)$.   ▯

Exercises

1. Let I be an ideal of L. Then each member of the derived series or descending central series of I is also an ideal of L.
2. Prove that L is solvable if and only if there exists a chain of subalgebras $L = L_0 \supset L_1 \supset L_2 \supset \ldots \supset L_k = 0$ such that L_{i+1} is an ideal of L_i and such that each quotient L_i/L_{i+1} is abelian.
3. Let char $\mathsf{F} = 2$. Prove that $\mathfrak{sl}(2, \mathsf{F})$ is nilpotent.
4. Prove that L is solvable (resp. nilpotent) if and only if ad L is solvable (resp. nilpotent)
5. Prove that the nonabelian two dimensional algebra constructed in (1.4) is solvable but not nilpotent. Do the same for the algebra in Exercise 1.2.
6. Prove that the sum of two nilpotent ideals of a Lie algebra L is again a nilpotent ideal. Therefore, L possesses a unique maximal nilpotent ideal. Determine this ideal for each algebra in Exercise 5.
7. Let L be nilpotent, K a proper subalgebra of L. Prove that $N_L(K)$ includes K properly.
8. Let L be nilpotent. Prove that L has an ideal of codimension 1.
9. Prove that every nilpotent Lie algebra L has an outer derivation (see (1.3)), as follows: Write $L = K + \mathsf{F}x$ for some ideal K of codimension one (Exercise 8). Then $C_L(K) \neq 0$ (why?). Choose n so that $C_L(K) \subset L^n$, $C_L(K) \not\subset L^{n+1}$, and let $z \in C_L(K) - L^{n+1}$. Then the linear map δ sending K to 0, x to z, is an outer derivation.
10. Let L be a Lie algebra, K an ideal of L such that L/K is nilpotent and such that ad $x|_K$ is nilpotent for all $x \in L$. Prove that L is nilpotent.

Chapter II

Semisimple Lie Algebras

In Chapter I we looked at Lie algebras over an arbitrary field F. Apart from introducing the basic notions and examples, we were able to prove only one substantial theorem (Engel's Theorem). Virtually all of the remaining theory to be developed in this book will require the assumption that F have characteristic 0. (Some of the exercises will indicate how counter-examples arise in prime characteristic.) Moreover, in order to have available the eigenvalues of ad x for arbitrary x (not just for ad x nilpotent), we shall assume that F is algebraically closed, except where otherwise specified. It is possible to work with a slightly less restrictive assumption on F (cf. Jacobson [1], p. 107), but we shall not do so here.

4. Theorems of Lie and Cartan

4.1. Lie's Theorem

The essence of Engel's Theorem for nilpotent Lie algebras is the existence of a common eigenvector for a Lie algebra consisting of nilpotent endomorphisms (Theorem 3.3). The next theorem is similar in nature, but requires algebraic closure, in order to assure that F will contain all required eigenvalues. It turns out to be necessary also to have char F $= 0$ (Exercise 3).

Theorem. *Let L be a solvable subalgebra of $\mathfrak{gl}(V)$, V finite dimensional. If $V \neq 0$, then V contains a common eigenvector for all the endomorphisms in L.*

Proof. Use induction on dim L, the case dim $L = 0$ being trivial. We attempt to imitate the proof of Theorem 3.3 (which the reader should review at this point). The idea is (1) to locate an ideal K of codimension one, (2) to show by induction that common eigenvectors exist for K, (3) to verify that L stabilizes a space consisting of such eigenvectors, and (4) to find in that space an eigenvector for a single $z \in L$ satisfying $L = K + \mathsf{F}z$.

Step (1) is easy. Since L is solvable, of positive dimension, L properly includes $[LL]$. $L/[LL]$ being abelian, any subspace is automatically an ideal. Take a subspace of codimension one, then its inverse image K is an ideal of codimension one in L (including $[LL]$).

For step (2), use induction to find a common eigenvector $v \in V$ for K (K is of course solvable; if $K = 0$, then L is abelian of dimension 1 and an eigenvector for a basis vector of L finishes the proof.) This means that for $x \in K$

$x.v = \lambda(x)v$, $\lambda: K \to$ F some linear function. Fix this λ, and denote by W the subspace

$$\{w \in V | x.w = \lambda(x)w, \text{ for all } x \in K\}; \text{ so } W \neq 0.$$

Step (3) consists in showing that L leaves W invariant. Assuming for the moment that this is done, proceed to step (4): Write $L = K + $Fz, and use the fact that F is algebraically closed to find an eigenvector $v_0 \in W$ of z (for some eigenvalue of z). Then v_0 is obviously a common eigenvector for L (and λ can be extended to a linear function on L such that $x.v_0 = \lambda(x)v_0$, $x \in L$).

It remains to show that L stabilizes W. Let $w \in W$, $x \in L$. To test whether or not $x.w$ lies in W, we must take arbitrary $y \in K$ and examine $yx.w = xy.w - [x, y].w = \lambda(y)x.w - \lambda([x, y])w$. Thus *we have to prove that* $\lambda([x, y]) = 0$. For this, fix $w \in W$, $x \in L$. Let $n > 0$ be the smallest integer for which w, $x.w, \ldots, x^n.w$ are linearly dependent. Let W_i be the subspace of W spanned by $w, x.w, \ldots, x^{i-1}.w$ (set $W_0 = 0$), so dim $W_n = n$, $W_n = W_{n+i}$ ($i \geq 0$) and x maps W_n into W_n. It is easy to check that each $y \in K$ leaves each W_i invariant. Relative to the basis $w, x.w, \ldots, x^{n-1}.w$ of W_n, we claim that $y \in K$ is represented by an upper triangular matrix whose diagonal entries equal $\lambda(y)$. This follows immediately from the congruence:

(*) $yx^i.w \equiv \lambda(y)x^i.w \pmod{W_i}$,

which we prove by induction on i, the case $i = 0$ being obvious. Write $yx^i.w = yxx^{i-1}.w = xyx^{i-1}.w - [x, y]x^{i-1}.w$. By induction, $yx^{i-1}.w = \lambda(y)x^{i-1}.w + w'$ ($w' \in W_{i-1}$); since x maps W_{i-1} into W_i (by construction), (*) therefore holds for all i.

According to our description of the way in which $y \in K$ acts on W_n, $Tr_{W_n}(y) = n\lambda(y)$. In particular, this is true for elements of K of the special form $[x, y]$ (x as above, y in K). But x, y both stabilize W_n, so $[x, y]$ acts on W_n as the *commutator* of two endomorphisms of W_n; its trace is therefore 0. We conclude that $n\lambda([x, y]) = 0$. Since char F $= 0$, this forces $\lambda([x, y]) = 0$, as required. ☐

Corollary A (Lie's Theorem). *Let L be a solvable subalgebra of* $\mathfrak{gl}(V)$, dim $V = n < \infty$. *Then L stabilizes some flag in V (in other words, the matrices of L relative to a suitable basis of V are upper triangular).*

Proof. Use the theorem, along with induction on dim V. ☐

More generally, let L be any solvable Lie algebra, $\phi: L \to \mathfrak{gl}(V)$ a finite dimensional representation of L. Then $\phi(L)$ is solvable, by Proposition 3.1(a), hence stabilizes a flag (Corollary A). For example, if ϕ is the adjoint representation, a flag of subspaces stable under L is just a chain of ideals of L, each of codimension one in the next. This proves:

Corollary B. *Let L be solvable. Then there exists a chain of ideals of L,* $0 = L_0 \subset L_1 \subset \ldots \subset L_n = L$, *such that dim $L_i = i$.* ☐

Corollary C. *Let L be solvable. Then $x \in [LL]$ implies that $ad_L x$ is nilpotent. In particular, $[LL]$ is nilpotent.*

Proof. Find a flag of ideals as in Corollary B. Relative to a basis $(x_1, \ldots,$ $x_n)$ of L for which $(x_1, \ldots, x_i)$ spans L_i, the matrices of ad L lie in $\mathfrak{t}(n, \mathsf{F})$. Therefore the matrices of [ad L, ad L] $=$ ad$_L$ [LL] lie in $\mathfrak{n}(n, \mathsf{F})$, the derived algebra of $\mathfrak{t}(n, \mathsf{F})$. It follows that ad$_L$ x is nilpotent for $x \in [LL]$; a fortiori ad$_{[LL]}$ x is nilpotent, so [LL] is nilpotent by Engel's Theorem. $\square$

4.2. Jordan-Chevalley decomposition

In this subsection only, char F *may be arbitrary.* We digress in order to introduce a very useful tool for the study of linear transformations. The reader may recall that the Jordan canonical form for a single endomorphism x over an algebraically closed field amounts to an expression of x in matrix form as a sum of blocks

$$\begin{bmatrix} a & 1 & & & 0 \\ & a & 1 & & \\ & & \cdot & \cdot & \\ & & & \cdot & \cdot \\ & & & & \cdot & 1 \\ 0 & & & & & a \end{bmatrix}$$

Since diag $(a, \ldots, a)$ commutes with the nilpotent matrix having one's just above the diagonal and zeros elsewhere, x is the sum of a diagonal and a nilpotent matrix which commute. We can make this decomposition more precise, as follows.

Call $x \in$ End V (V finite dimensional) **semisimple** if the roots of its minimal polynomial over F are all distinct. Equivalently (F being algebraically closed), x is semisimple if and only if x is diagonalizable. We remark that two commuting semisimple endomorphisms can be simultaneously diagonalized; therefore, their sum or difference is again semisimple (Exercise 2). Also, if x is semisimple and maps a subspace W of V into itself, then obviously the restriction of x to W is semisimple.

Proposition. *Let V be a finite dimensional vector space over* F, $x \in$ *End V.*

(a) There exist unique x_s, $x_n \in$ End V satisfying the conditions: $x = x_s + x_n$, x_s is semisimple, x_n is nilpotent, x_s and x_n commute.

(b) There exist polynomials $p(T)$, $q(T)$ in one indeterminate, without constant term, such that $x_s = p(x)$, $x_n = q(x)$. In particular, x_s and x_n commute with any endomorphism commuting with x.

(c) If $A \subset B \subset V$ are subspaces, and x maps B into A, then x_s and x_n also map B into A.

The decomposition $x = x_s + x_n$ is called the (additive) **Jordan-Chevalley decomposition** of x, or just the Jordan decomposition; x_s, x_n are called (respectively) the **semisimple part** and the **nilpotent part** of x.

Proof. Let $a_1, \ldots, a_k$ (with multiplicities $m_1, \ldots, m_k$) be the distinct eigenvalues of x, so the characteristic polynomial is $\Pi(T - a_i)^{m_i}$. If $V_i =$ Ker $(x - a_i \cdot 1)^{m_i}$, then V is the direct sum of the subspaces $V_1, \ldots, V_k$, each stable under x. On V_i, x clearly has characteristic polynomial $(T-a_i)^{m_i}$. Now apply the Chinese Remainder Theorem (for the ring $F[T]$) to locate a polynomial $p(T)$ satisfying the congruences, with pairwise relatively prime moduli: $p(T) \equiv a_i \pmod{(T-a_i)^{m_i}}$, $p(T) \equiv 0 \pmod{T}$.) (Notice that the last congruence is superfluous if 0 is an eigenvalue of x, while otherwise T is relatively prime to the other moduli.) Set $q(T) = T-p(T)$. Evidently each of $p(T)$, $q(T)$ has zero constant term, since $p(T) \equiv 0 \pmod{T}$.

Set $x_s = p(x)$, $x_n = q(x)$. Since they are polynomials in x, x_s and x_n commute with each other, as well as with all endomorphisms which commute with x. They also stabilize all subspaces of V stabilized by x, in particular the V_i. The congruence $p(T) \equiv a_i \pmod{(T-a_i)^{m_i}}$ shows that the restriction of $x_s - a_i \cdot 1$ to V_i is zero for all i, hence that x_s acts diagonally on V_i with single eigenvalue a_i. By definition, $x_n = x - x_s$, which makes it clear that x_n is nilpotent. Because $p(T)$, $q(T)$ have no constant term, (c) is also obvious at this point.

It remains only to prove the uniqueness assertion in (a). Let $x = s+n$ be another such decomposition, so we have $x_s - s = n - x_n$. Because of (b), all endomorphisms in sight commute. Sums of commuting semisimple (resp. nilpotent) endomorphisms are again semisimple (resp. nilpotent), whereas only 0 can be both semisimple and nilpotent. This forces $s = x_s$, $n = x_n$. $\square$

To indicate why the Jordan decomposition will be a valuable tool, we look at a special case. Consider the adjoint representation of the Lie algebra $\mathfrak{gl}(V)$, V finite dimensional. If $x \in \mathfrak{gl}(V)$ is nilpotent, then so is ad x (Lemma 3.2). Similarly, *if x is semisimple, then so is ad x.* We verify this as follows. Choose a basis $(v_1, \ldots, v_n)$ of V relative to which x has matrix diag $(a_1, \ldots, a_n)$. Let $\{e_{ij}\}$ be the standard basis of $\mathfrak{gl}(V)$ (1.2) relative to $(v_1, \ldots, v_n)$: $e_{ij}(v_k) = \delta_{jk}v_i$. Then a quick calculation (see formula (*) in (1.2)) shows that ad $x(e_{ij}) = (a_i - a_j)e_{ij}$. So ad x has diagonal matrix, relative to the chosen basis of $\mathfrak{gl}(V)$.

Lemma A. *Let $x \in \text{End } V$ (dim $V < \infty$), $x = x_s + x_n$ its Jordan decomposition. Then ad $x = \text{ad } x_s + \text{ad } x_n$ is the Jordan decomposition of ad x (in End (End V)).*

Proof. We have seen that ad x_s, ad x_n are respectively semisimple, nilpotent; they commute, since [ad x_s, ad x_n] = ad [x_s, x_n] = 0. Then part (a) of the proposition applies. $\square$

A further useful fact is the following.

Lemma B. *Let $\mathfrak{A}$ be a finite dimensional F-algebra. Then Der $\mathfrak{A}$ contains the semisimple and nilpotent parts (in End $\mathfrak{A}$) of all its elements.*

Proof. If $\delta \in \text{Der } \mathfrak{A}$, let σ, $\nu \in \text{End } \mathfrak{A}$ be its semisimple and nilpotent parts, respectively. It will be enough to show that $\sigma \in \text{Der } \mathfrak{A}$. If $a \in F$, set $\mathfrak{A}_a = \{x \in \mathfrak{A} | (\delta - a.1)^k x = 0 \text{ for some } k \text{ (depending on } x)\}$. Then $\mathfrak{A}$ is the direct sum of those $\mathfrak{A}_a$ for which a is an eigenvalue of δ (or σ), and σ acts on $\mathfrak{A}_a$ as scalar multiplication by a. We can verify, for arbitrary a, $b \in F$, that $\mathfrak{A}_a \mathfrak{A}_b \subset \mathfrak{A}_{a+b}$, by means of the general formula: (*) $(\delta - (a+b).1)^n(xy)$

$$= \sum_{i=0}^{n} \binom{n}{i} \, ((\delta - a.1)^{n-i} x) \cdot ((\delta - b.1)^{i} y), \quad \text{for } x, y \in \mathfrak{A}.$$ (This formula is easily checked by induction on n.) Now if $x \in \mathfrak{A}_a$, $y \in \mathfrak{A}_b$, then $\sigma(xy) = (a+b)xy$, because $xy \in \mathfrak{A}_{a+b}$ (possibly equal to 0); on the other hand, $(\sigma x)y + x(\sigma y) = (a+b)xy$. By directness of the sum $\mathfrak{A} = \coprod \mathfrak{A}_a$, it follows that σ is a derivation, as required. □

4.3. Cartan's Criterion

We are now ready to obtain a powerful criterion for solvability of a Lie algebra L, based on the traces of certain endomorphisms of L. It is obvious that L will be solvable if $[LL]$ is nilpotent (this is the converse of Corollary 4.1C). In turn, Engel's Theorem says that $[LL]$ will be nilpotent if (and only if) each $\mathrm{ad}_{[LL]} x$, $x \in [LL]$, is nilpotent. We begin, therefore, with a "trace" criterion for nilpotence of an endomorphism.

Lemma. *Let A, B be two subspaces of $\mathfrak{gl}(V)$, $\dim V < \infty$. Set $M = \{x \in \mathfrak{gl}(V) | [x, B] \subset A\}$. Suppose $x \in M$ satisfies $Tr(xy) = 0$ for all $y \in M$. Then x is nilpotent.*

Proof. Let $x = s + n$ ($s = x_s$, $n = x_n$) be the Jordan decomposition of x. Fix a basis $v_1, \ldots, v_m$ of V relative to which s has matrix $\mathrm{diag}(a_1, \ldots, a_m)$. Let E be the vector subspace of F (over the prime field $\mathbf{Q}$) spanned by the eigenvalues $a_1, \ldots, a_m$. We have to show that $s = 0$, or equivalently, that $E = 0$. Since E has finite dimension over $\mathbf{Q}$ (by construction), it will suffice to show that the dual space E^* is 0, i.e., that any linear function $f : E \to \mathbf{Q}$ is zero.

Given f, let y be that element of $\mathfrak{gl}(V)$ whose matrix relative to our given basis is $\mathrm{diag}(f(a_1), \ldots, f(a_m))$. If $\{e_{ij}\}$ is the corresponding basis of $\mathfrak{gl}(V)$, we saw in (4.2) that: $\mathrm{ad}\, s(e_{ij}) = (a_i - a_j)e_{ij}$, $\mathrm{ad}\, y(e_{ij}) = (f(a_i) - f(a_j))e_{ij}$. Now let $r(T) \in \mathsf{F}[T]$ be a polynomial without constant term satisfying $r(a_i - a_j) = f(a_i) - f(a_j)$ for all pairs i, j. The existence of such $r(T)$ follows from Lagrange interpolation; there is no ambiguity in the assigned values, since $a_i - a_j = a_k - a_l$ implies (by linearity of f) that $f(a_i) - f(a_j) = f(a_k) - f(a_l)$. Evidently $\mathrm{ad}\, y = r\,(\mathrm{ad}\, s)$.

Now $\mathrm{ad}\, s$ is the semisimple part of $\mathrm{ad}\, x$, by Lemma A of (4.2), so it can be written as a polynomial in $\mathrm{ad}\, x$ without constant term (Proposition 4.2). Therefore, $\mathrm{ad}\, y$ is also a polynomial in $\mathrm{ad}\, x$ without constant term. By hypothesis, $\mathrm{ad}\, x$ maps B into A, so we also have $\mathrm{ad}\, y\,(B) \subset A$, i.e., $y \in M$. Using the hypothesis of the lemma, $Tr(xy) = 0$, we get $\Sigma a_i f(a_i) = 0$. The left side is a $\mathbf{Q}$-linear combination of elements of E; applying f, we obtain $\Sigma f(a_i)^2 = 0$. But the numbers $f(a_i)$ are rational, so this forces all of them to be 0. Finally, f must be identically 0, because the a_i span E. □

Before stating our solvability criterion, we record a useful identity: If x, y, z are endomorphisms of a finite dimensional vector space, then (*) $Tr([x, y]z) = Tr(x[y, z])$. To verify this, write $[x, y]z = xyz - yxz$, $x[y, z] = xyz - xzy$, and use the fact that $Tr(y(xz)) = Tr((xz)y)$.

Theorem (Cartan's Criterion). *Let* L *be a subalgebra of* $\mathfrak{gl}(V)$, V *finite dimensional. Suppose that* $Tr(xy) = 0$ *for all* $x \in [LL]$, $y \in L$. *Then* L *is solvable.*

Proof. As remarked at the beginning of (4.3), it will suffice to prove that $[LL]$ is nilpotent, or just that all x in $[LL]$ are nilpotent endomorphisms (Lemma 3.2 and Engel's Theorem). For this we apply the above lemma to the situation: V as given, $A = [LL]$, $B = L$, so $M = \{x \in \mathfrak{gl}(V) | [x, L] \subset [LL]\}$. Obviously $L \subset M$. Our hypothesis is that $Tr(xy) = 0$ for $x \in [LL]$, $y \in L$, whereas to conclude from the lemma that each $x \in [LL]$ is nilpotent we need the stronger statement: $Tr(xy) = 0$ for $x \in [LL]$, $y \in M$.

Now if $[x, y]$ is a typical generator of $[LL]$, and if $z \in M$, then identity (*) above shows that $Tr([x, y]z) = Tr(x[y, z]) = Tr([y, z]x)$. By definition of M, $[y, z] \in [LL]$, so the right side is 0 by hypothesis. $\square$

Corollary. *Let* L *be a Lie algebra such that* $Tr(\text{ad } x \text{ ad } y) = 0$ *for all* $x \in [LL]$, $y \in L$. *Then* L *is solvable.*

Proof. Applying the theorem to the adjoint representation of L, we get ad L solvable. Since Ker ad $= Z(L)$ is solvable, L itself is solvable (Proposition 3.1). $\square$

Exercises

1. Let $L \subset \mathfrak{gl}(V)$ be a classical linear Lie algebra (1.2). Use Lie's Theorem, along with Exercise 1.12, to prove that Rad $L = Z(L)$. In particular, $\mathfrak{sl}(n, \mathsf{F})$ is semisimple (cf. Exercise 2.3). [Observe that Rad L lies in each maximal solvable subalgebra B of L. Select a basis of V so that $B = L \cap \mathfrak{t}(n, \mathsf{F})$, and notice that the transpose of B is also a maximal solvable subalgebra of L. Conclude that Rad $L \subset L \cap \mathfrak{d}(n, \mathsf{F})$, then that Rad $L = Z(L)$.]

2. Show that the proof of Theorem 4.1 still goes through in prime characteristic, provided dim V is less than char F.

3. This exercise illustrates the failure of Lie's Theorem when F is allowed to have prime characteristic p. Consider the $p \times p$ matrices:

$$x = \begin{bmatrix} 0 & 1 & 0 & . & . & . & 0 \\ 0 & 0 & 1 & 0 & . & . & 0 \\ . & . & . & . & . & . \\ 0 & . & . & . & . & 1 \\ 1 & . & . & . & . & 0 \end{bmatrix}, \quad y = \text{diag}(0, 1, 2, 3, \ldots, p-1).$$

Check that $[x, y] = x$, hence that x and y span a two dimensional solvable subalgebra L of $\mathfrak{gl}(p, \mathsf{F})$. Verify that x, y have no common eigenvector.

4. When $p = 2$, Exercise 3.3 shows that a solvable Lie algebra of endomorphisms over a field of prime characteristic p need not have derived

algebra consisting of nilpotent endomorphisms (cf. Corollary C of Theorem 4.1). For arbitrary p, construct a counterexample to Corollary C as follows: Start with $L \subset \mathfrak{gl}(p, \mathsf{F})$ as in Exercise 3. Form the vector space direct sum $M = L + \mathsf{F}^p$, and make M a Lie algebra by decreeing that F^p is abelian, while L has its usual product and acts on F^p in the given way. Verify that M is solvable, but that its derived algebra ($= \mathsf{F}x + \mathsf{F}^p$) fails to be nilpotent.

5. If $x, y \in \mathrm{End}\ V$ commute, prove that $(x+y)_s = x_s + y_s$, and $(x+y)_n = x_n + y_n$. Show by example that this can fail if x, y fail to commute. [Show first that x, y semisimple (resp. nilpotent) implies $x + y$ semisimple (resp. nilpotent).]

6. Check formula (*) at the end of (4.2).

7. Prove the converse of Theorem 4.3.

8. Note that it suffices to check the hypothesis of Theorem 4.3 (or its corollary) for x, y ranging over a basis of L. For the example given in Exercise 1.2, verify solvability by using Cartan's Criterion.

Notes

The proofs here follow Serre [1]. The systematic use of the Jordan decomposition in linear algebraic groups originates with Chevalley [1]; see also Borel [1], where the additive Jordan decomposition in the Lie algebra is emphasized.

5. Killing form

5.1. Criterion for semisimplicity

Let L be any Lie algebra. If $x, y \in L$, define $\kappa(x, y) = Tr(\mathrm{ad}\ x\ \mathrm{ad}\ y)$. Then κ is a symmetric bilinear form on L, called the **Killing form**. κ is also **associative**, in the sense that $\kappa([xy], z) = \kappa(x, [yz])$. This follows from the identity recorded in (4.3): $Tr([x, y]z) = Tr(x[y, z])$, for endomorphisms x, y, z of a finite dimensional vector space.

The following lemma will be handy later on.

Lemma. *Let I be an ideal of L. If κ is the Killing form of L and κ_I the Killing form of I (viewed as Lie algebra), then $\kappa_I = \kappa|_{I \times I}$.*

Proof. First, a simple fact from linear algebra: If W is a subspace of a (finite dimensional) vector space V, and ϕ an endomorphism of V mapping V into W, then $Tr\phi = Tr(\phi|_W)$. (To see this, extend a basis of W to a basis of V and look at the resulting matrix of ϕ.) Now if $x, y \in I$, then $(\mathrm{ad}\ x)(\mathrm{ad}\ y)$ is an endomorphism of L, mapping L into I, so its trace $\kappa(x, y)$ coincides with the trace $\kappa_I(x, y)$ of $(\mathrm{ad}\ x)(\mathrm{ad}\ y)|_I = (\mathrm{ad}_I\ x)(\mathrm{ad}_I\ y)$. $\quad\square$

In general, a symmetric bilinear form $\beta(x, y)$ is called **nondegenerate** if its **radical** S is 0, where $S = \{x \in L | \beta(x, y) = 0 \text{ for all } y \in L\}$. Because the Killing form is associative, its radical is more than just a subspace: S is an *ideal* of L. From linear algebra, a practical way to test nondegeneracy is as follows: Fix a basis $x_1, \ldots, x_n$ of L. Then κ is nondegenerate if and only if the $n \times n$ matrix whose i, j entry is $\kappa(x_i, x_j)$ has nonzero determinant.

As an example, we compute the Killing form of $\mathfrak{sl}(2, \mathsf{F})$, using the standard basis (Example 2.1), which we write in the order (x, h, y). The matrices become:

$$\text{ad } h = \text{diag } (2, 0, -2), \quad \text{ad } x = \begin{pmatrix} 0 & -2 & 0 \\ 0 & 0 & 1 \\ 0 & 0 & 0 \end{pmatrix}, \quad \text{ad } y = \begin{pmatrix} 0 & 0 & 0 \\ -1 & 0 & 0 \\ 0 & 2 & 0 \end{pmatrix}.$$

Therefore κ has matrix $\begin{pmatrix} 0 & 0 & 4 \\ 0 & 8 & 0 \\ 4 & 0 & 0 \end{pmatrix}$, with determinant -128, and κ is non-degenerate. (This is still true so long as char $\mathsf{F} \neq 2$.)

Recall that a (nonzero) Lie algebra L is called semisimple if Rad $L = 0$. This is equivalent to requiring that L have no nonzero abelian ideals: indeed, any such ideal must be in the radical, and conversely, the radical (if nonzero) includes such an ideal of L, viz., the last nonzero term in the derived series of Rad L (cf. Exercise 3.1).

Theorem. *Let L be a (nonzero) Lie algebra. Then L is semisimple if and only if its Killing form is nondegenerate.*

Proof. Suppose first that Rad $L = 0$. Let S be the radical of κ. By definition, $Tr(\text{ad } x \text{ ad } y) = 0$ for all $x \in S$, $y \in L$ (in particular, for $y \in [SS]$). According to Cartan's Criterion (4.3), $\text{ad}_L S$ is solvable, hence S is solvable. But we remarked above that S is an ideal of L, so $S \subset$ Rad $L = 0$, and κ is nondegenerate.

Conversely, let $S = 0$. To prove that L is semisimple, it will suffice to prove that every abelian ideal I of L is included in S. Suppose $x \in I$, $y \in L$. Then ad x ad y maps $L \to L \to I$, and $(\text{ad } x \text{ ad } y)^2$ maps L into $[II] = 0$. This means that ad x ad y is nilpotent, hence that $0 = Tr(\text{ad } x \text{ ad } y) = \kappa(x, y)$, so $I \subset S = 0$. (This half of the proof remains valid even in prime characteristic (Exercise 6).) □

The proof shows that we always have $S \subset$ Rad L; however, the reverse inclusion need not hold (Exercise 4).

5.2. Simple ideals of L

First a definition. A Lie algebra L is said to be the **direct sum** of ideals $I_1, \ldots, I_t$ provided $L = I_1 + \ldots + I_t$ (direct sum of subspaces). This condition forces $[I_i I_j] \subset I_i \cap I_j = 0$ if $i \neq j$ (so the algebra L can be viewed as gotten from the Lie algebras I_i by defining Lie products componentwise for the external direct sum of these as vector spaces). We write $L = I_1 \oplus \ldots \oplus I_t$.

Theorem. *Let L be semisimple. Then there exist ideals $L_1, \ldots, L_t$ of L which are simple (as Lie algebras), such that $L = L_1 \oplus \ldots \oplus L_t$. Every simple ideal of L coincides with one of the L_i. Moreover, the Killing form of L_i is the restriction of κ to $L_i \times L_i$.*

Proof. As a first step, let I be an arbitrary ideal of L. Then $I^\perp = \{x \in L \mid \kappa(x, y) = 0$ for all $y \in I\}$ is also an ideal, by the associativity of κ. Cartan's Criterion, applied to the Lie algebra I, shows that the ideal $I \cap I^\perp$ of L is solvable (hence 0). Therefore, since $\dim I + \dim I^\perp = \dim L$, we must have $L = I \oplus I^\perp$.

Now proceed by induction on $\dim L$ to obtain the desired decomposition into direct sum of simple ideals. If L has no nonzero proper ideal, then L is simple already and we're done. Otherwise let L_1 be a minimal nonzero ideal; by the preceding paragraph, $L = L_1 \oplus L_1^\perp$. In particular, any ideal of L_1 is also an ideal of L, so L_1 is semisimple (hence simple, by minimality). For the same reason, $L_1^\perp$ is semisimple; by induction, it splits into a direct sum of simple ideals, which are also ideals of L. The decomposition of L follows.

Next we have to prove that these simple ideals are unique. If I is any simple ideal of L, then $[IL]$ is also an ideal of I, nonzero because $Z(L) = 0$; this forces $[IL] = I$. On the other hand, $[IL] = [IL_1] \oplus \ldots \oplus [IL_t]$, so all but one summand must be 0. Say $[IL_i] = I$. Then $I \subset L_i$, and $I = L_i$ (because L_i is simple).

The last assertion of the theorem follows from Lemma 5.1.   ☐

Corollary. *If L is semisimple, then $L = [LL]$, and all ideals and homomorphic images of L are semisimple (or 0). Moreover, each ideal of L is a sum of certain simple ideals of L.*   ☐

5.3. *Inner derivations*

There is a further important consequence of nondegeneracy of the Killing form. Before stating it we recall explicitly the result of Exercise 2.1: $\text{ad } L$ is an ideal in $\text{Der } L$ (for any Lie algebra L). The proof depends on the simple observation: (*) $[\delta, \text{ad } x] = \text{ad } (\delta x)$, $x \in L$, $\delta \in \text{Der } L$.

Theorem. *If L is semisimple, then $\text{ad } L = \text{Der } L$ (i.e., every derivation of L is inner).*

Proof. Since L is semisimple, $Z(L) = 0$. Therefore, $L \to \text{ad } L$ is an isomorphism of Lie algebras. In particular, $M = \text{ad } L$ itself has nondegenerate Killing form (Theorem 5.1). If $D = \text{Der } L$, we just remarked that $[D, M] \subset M$. This implies (by Lemma 5.1) that κ_M is the restriction to $M \times M$ of the Killing form κ_D of D. In particular, if $I = M^\perp$ is the subspace of D orthogonal to M under κ_D, then the nondegeneracy of κ_M forces $I \cap M = 0$. Both I and M are ideals of D, so we obtain $[I, M] = 0$. If $\delta \in I$, this forces $\text{ad } (\delta x) = 0$ for all $x \in L$ (by (*)), so in turn $\delta x = 0$ $(x \in L)$ because ad is $1-1$, and $\delta = 0$. Conclusion: $I = 0$, $\text{Der } L = M = \text{ad } L$.   ☐

5.4. *Abstract Jordan decomposition*

Theorem 5.3 can be used to introduce an abstract Jordan decomposition in an arbitrary semisimple Lie algebra L. Recall (Lemma B of (4.2)) that if $\mathfrak{A}$ is any F-algebra of finite dimension, then Der $\mathfrak{A}$ contains the semisimple and nilpotent parts in End $\mathfrak{A}$ of all its elements. In particular, since Der L coincides with ad L (5.3), while $L \to$ ad L is $1-1$, each $x \in L$ determines unique elements s, $n \in L$ such that ad $x =$ ad $s+$ad n is the usual Jordan decomposition of ad x (in End L). This means that $x = s+n$, with $[sn] = 0$, s **ad-semisimple** (i.e., ad s semisimple), n ad-nilpotent. We write $s = x_s$, $n = x_n$, and (by abuse of language) call these the **semisimple** and **nilpotent parts** of x.

The alert reader will object at this point that the notation x_s, x_n is ambiguous in case L happens to be a linear Lie algebra. It will be shown in (6.4) that the abstract decomposition of x just obtained does in fact agree with the usual Jordan decomposition in all such cases. For the moment we shall be content to point out that this is true in the special case $L = \mathfrak{sl}(V)$ (V finite dimensional): Write $x = x_s+x_n$ in End V (usual Jordan decomposition), $x \in L$. Since x_n is a nilpotent endomorphism, its trace is 0 and therefore $x_n \in L$. This forces x_s also to have trace 0, so $x_s \in L$. Moreover, $\mathrm{ad}_{\mathfrak{sl}(V)}x_s$ is semisimple (Lemma A of (4.2)), so $\mathrm{ad}_L\, x_s$ is a fortiori semisimple; similarly $\mathrm{ad}_L\, x_n$ is nilpotent, and $[\mathrm{ad}_L\, x_s, \mathrm{ad}_L\, x_n] = \mathrm{ad}_L[x_s x_n] = 0$. By the uniqueness of the abstract Jordan decomposition in L, $x = x_s+x_n$ must be it.

Exercises

1. Prove that if L is nilpotent, the Killing form of L is identically zero.
2. Prove that L is solvable if and only if $[LL]$ lies in the radical of the Killing form.
3. Let L be the two dimensional nonabelian Lie algebra (1.4), which is solvable. Prove that L has nontrivial Killing form.
4. Let L be the three dimensional solvable Lie algebra of Exercise 1.2. Compute the radical of its Killing form.
5. Let $L = \mathfrak{sl}(2, \mathsf{F})$. Compute the basis of L dual to the standard basis, relative to the Killing form.
6. Let char $\mathsf{F} = p \neq 0$. Prove that L is semisimple if its Killing form is nondegenerate. Show by example that the converse fails. [Look at $\mathfrak{sl}(3, \mathsf{F})$ modulo its center, when char $\mathsf{F} = 3$.]
7. Relative to the standard basis of $\mathfrak{sl}(3, \mathsf{F})$, compute the determinant of κ. Which primes divide it?
8. Let $L = L_1 \oplus \ldots \oplus L_t$ be the decomposition of a semisimple Lie algebra L into its simple ideals. Show that the semisimple and nilpotent parts of $x \in L$ are the sums of the semisimple and nilpotent parts in the various L_i of the components of x.

Notes

Even in prime characteristic, nondegeneracy of the Killing form has very strong implications for the structure of a Lie algebra. See Seligman [1], Pollack [1], Kaplansky [1].

6. Complete reducibility of representations

In this section all representations are finite dimensional, unless otherwise noted.

We are going to study a semisimple Lie algebra L by means of its adjoint representation (see §8). It turns out that L is built up from copies of $\mathfrak{sl}(2, F)$; to study the adjoint action of such a three dimensional subalgebra of L, we need precise information about the representations of $\mathfrak{sl}(2, F)$, to be given in §7 below. First we prove an important general theorem (due to Weyl) about representations of an arbitrary semisimple Lie algebra.

6.1. Modules

Let L be a Lie algebra. It is often convenient to use the language of modules along with the (equivalent) language of representations. As in other algebraic theories, there is a natural definition. A vector space V, endowed with an operation $L \times V \to V$ (denoted $(x, v) \mapsto x.v$ or just xv) is called an **L-module** if the following conditions are satisfied:

(M1) $(ax+by).v = a(x.v)+b(y.v)$,

(M2) $x.(av+bw) = a(x.v)+b(x.w)$,

(M3) $[xy].v = x.y.v - y.x.v.$ $(x, y \in L; v, w \in V; a, b \in F).$

For example, if $\phi: L \to \mathfrak{gl}(V)$ is a representation of L, then V may be viewed as an L-module via the action $x.v = \phi(x)(v)$. Conversely, given an L-module V, this equation defines a representation $\phi: L \to \mathfrak{gl}(V)$.

A **homomorphism of L-modules** is a linear map $\phi: V \to W$ such that $\phi(x.v) = x.\phi(v)$. The kernel of such a homomorphism is then an L-submodule of V (and the standard homomorphism theorems all go through without difficulty). When ϕ is an isomorphism of vector spaces, we call it an **isomorphism** of L-modules; in this case, the two modules are said to afford **equivalent** representations of L. An L-module V is called **irreducible** if it has precisely two L-submodules (itself and 0); in particular, *we do not regard a zero dimensional vector space as an irreducible L-module*. We do, however, allow a one dimensional space on which L acts (perhaps trivially) to be called irreducible. V is called **completely reducible** if V is a direct sum of irreducible L-submodules, or equivalently (Exercise 2), if each L-submodule W of V has a complement W' (an L-submodule such that $V = W \oplus W'$). When

W, W' are arbitrary L-modules, we can of course make their direct sum an L-module in the obvious way, by defining $x.(w, w') = (x.w, x.w')$. These notions are all standard and also make sense when dim $V = \infty$. Of course, the terminology "irreducible" and "completely reducible" applies equally well to representations of L.

Given a representation $\phi \colon L \to \mathfrak{gl}(V)$, the associative algebra (with 1) generated by $\phi(L)$ in End V leaves invariant precisely the same subspaces as L. Therefore, all the usual results (e.g., Jordan-Hölder Theorem) for modules over associative rings hold for L as well. For later use, we recall the well known Schur's Lemma.

Schur's Lemma. *Let $\phi \colon L \to \mathfrak{gl}(V)$ be irreducible. Then the only endomorphisms of V commuting with all $\phi(x)$ $(x \in L)$ are the scalars.*   □

L itself is an L-module (for the adjoint representation). An L-submodule is just an ideal, so tt follows that a simple algebra L is irreducible as L-module, while a semisimple algebra is completely reducible (Theorem 5.2).

For later use we mention a couple of standard ways in which to fabricate new L-modules from old ones. Let V be an L-module. Then the dual vector space V^* becomes an L-module (called the **dual** or **contragredient**) if we define, for $f \in V^*$, $v \in V$, $x \in L$: $(x.f)(v) = -f(x.v)$. Axioms (M1), (M2) are almost obvious, so we just check (M3):

$$
\begin{aligned}
([xy].f)(v) &= -f([xy].v) \\
&= -f(x.y.v - y.x.v) \\
&= -f(x.y.v) + f(y.x.v) \\
&= (x.f)(y.v) - (y.f)(x.v) \\
&= -(y.x.f)(v) + (x.y.f)(v) \\
&= ((x.y - y.x).f)(v).
\end{aligned}
$$

If V, W are L-modules, let $V \otimes W$ be the tensor product over F of the underlying vector spaces. Recall that if V, W have respective bases $(v_1, \ldots, v_m)$ and $(w_1, \ldots, w_n)$, then $V \otimes W$ has a basis consisting of the mn vectors $v_i \otimes w_j$. The reader may know how to give a module structure to the tensor product of two modules for a group G: on the generators $v \otimes w$, require $g.(v \otimes w) = g.v \otimes g.w$. For Lie algebras the correct definition is gotten by "differentiating" this one: $x.(v \otimes w) = x.v \otimes w + v \otimes x.w$. As before, the crucial axiom to verify is (M3):

$$
\begin{aligned}
[xy].(v \otimes w) &= [xy].v \otimes w + v \otimes [xy].w \\
&= (x.y.v - y.x.v) \otimes w + v \otimes (x.y.w - y.x.w) \\
&= (x.y.v \otimes w + v \otimes x.y.w) - (y.x.v \otimes w + v \otimes y.x.w) \\
&= (x.y - y.x).(v \otimes w).
\end{aligned}
$$

Given a vector space V over F, there is a standard (and very useful) isomorphism of vector spaces: $V^* \otimes V \to$ End V, given by sending a typical generator $f \otimes v$ $(f \in V^*, v \in V)$ to the endomorphism whose value at $w \in V$

is $f(w)v$. It is a routine matter (using dual bases) to show that this does set up an epimorphism $V^* \otimes V \to \mathrm{End}\ V$; since both sides have dimension n^2 ($n = \dim V$), this must be an isomorphism.

Now if V (hence V^*) is in addition an L-module, then $V^* \otimes V$ becomes an L-module in the way described above. Therefore, End V can also be viewed as an L-module via the isomorphism just exhibited. This action of L on End V can also be described directly: $(x.f)(v) = x.f(v) - f(x.v)$, $x \in L$, $f \in \mathrm{End}\ V$, $v \in V$ (verify!). More generally, if V and W are two L-modules, then L acts naturally on the space Hom (V, W) of linear maps by the rule $(x.f)(v) = x.f(v) - f(x.v)$. (This action arises from the isomorphism between Hom (V, W) and $V^* \otimes W$.)

6.2. Casimir element of a representation

In §5 we used Cartan's trace criterion for solvability to prove that a semisimple Lie algebra L has nondegenerate Killing form. More generally, let L be semisimple and let $\phi: L \to \mathfrak{gl}(V)$ be a **faithful** (i.e., 1–1) representation of L. Define a symmetric bilinear form $\beta(x, y) = Tr(\phi(x)\phi(y))$ on L. The form β is associative, thanks to identity (*) in (4.3), so in particular its radical S is an ideal of L. Moreover, β is nondegenerate: indeed, Theorem 4.3 shows that $\phi(S) \cong S$ is solvable, so $S = 0$. (The Killing form is just β in the special case $\phi = \mathrm{ad}$.)

Now let L be semisimple, β any nondegenerate symmetric associative bilinear form on L. If $(x_1, \ldots, x_n)$ is a basis of L, there is a uniquely determined dual basis $(y_1, \ldots, y_n)$ relative to β, satisfying $\beta(x_i, y_j) = \delta_{ij}$. If $x \in L$, we can write $[xx_i] = \sum_j a_{ij}x_j$ and $[xy_i] = \sum_j b_{ij}y_j$. Using the associativity of β, we compute: $a_{ik} = \sum_j a_{ij}\beta(x_j, y_k) = \beta([xx_i], y_k) = \beta(-[x_ix], y_k) = \beta(x_i, -[xy_k]) = -\sum_j b_{kj}\beta(x_i, y_j) = -b_{ki}$.

If $\phi: L \to \mathfrak{gl}(V)$ is any representation of L, write $c_\phi(\beta) = \sum_i \phi(x_i)\phi(y_i) \in$ End V (x_i, y_i running over dual bases relative to β, as above). Using the identity (in End V) $[x, yz] = [x, y]z + y[x, z]$ and the fact that $a_{ik} = -b_{ki}$ (for $x \in L$ as above), we obtain: $[\phi(x), c_\phi(\beta)] = \sum_i [\phi(x), \phi(x_i)]\phi(y_i) + \sum_i \phi(x_i)[\phi(x), \phi(y_i)] = \sum_{i,j} a_{ij}\phi(x_j)\phi(y_i) + \sum_{i,j} b_{ij}\phi(x_i)\phi(y_j) = 0$. In other words, $c_\phi(\beta)$ *is an endomorphism of V commuting with $\phi(L)$.*

To bring together the preceding remarks, let $\phi: L \to \mathfrak{gl}(V)$ be a faithful representation, with (nondegenerate!) trace form $\beta(x, y) = Tr(\phi(x)\phi(y))$. In this case, having fixed a basis $(x_1, \ldots, x_n)$ of L, we write simply c_ϕ for $c_\phi(\beta)$ and call this the **Casimir element of ϕ**. Its trace is $\sum_i Tr(\phi(x_i)\phi(y_i)) = \sum_i \beta(x_i, y_i) = \dim L \neq 0$. In case ϕ is also irreducible, Schur's Lemma (6.1) implies that c_ϕ is a scalar (equal to $\dim L/\dim V$, in view of the preceding sentence); in this case we see that c_ϕ is independent of the basis of L which we chose.

Example. $L = \mathfrak{sl}(2, \mathsf{F})$, $V = \mathsf{F}^2$, ϕ the identity map $L \to \mathfrak{gl}(V)$. Let (x, h, y) be the standard basis of L (2.1). It is quickly seen that the dual basis relative to the trace form is $(y, 1/2h, x)$, so $c_\phi = xy + (1/2)h^2 + yx = \begin{pmatrix} 3/2 & 0 \\ 0 & 3/2 \end{pmatrix}$. Notice that $3/2 = \dim L/\dim V$.

When ϕ is no longer faithful, a slight modification is needed. Ker ϕ is an ideal of L, hence a sum of certain simple ideals (Corollary 5.2). Let L' denote the sum of the remaining simple ideals (Theorem 5.2). Then the restriction of ϕ to L' is a faithful representation of L', and we make the preceding construction (using dual bases of L'); the resulting element of End V is again called the Casimir element of ϕ and denoted c_ϕ. Evidently it commutes with $\phi(L) = \phi(L')$, etc.

One last remark: It is often convenient to assume that we are dealing with a faithful representation of L, which amounts to studying the representations of certain (semisimple) ideals of L. If L is simple, only the one dimensional module (on which L acts trivially) or the module 0 will fail to be faithful.

6.3. Weyl's Theorem

Lemma. *Let $\phi: L \to \mathfrak{gl}(V)$ be a representation of a semisimple Lie algebra L. Then $\phi(L) \subset \mathfrak{sl}(V)$. In particular, L acts trivially on any one dimensional L-module.*

Proof. Use the fact that $L = [LL]$ (5.2) along with the fact that $\mathfrak{sl}(V)$ is the derived algebra of $\mathfrak{gl}(V)$. ☐

Theorem (Weyl). *Let $\phi: L \to \mathfrak{gl}(V)$ be a (finite dimensional) representation of a semisimple Lie algebra, $V \neq 0$. Then ϕ is completely reducible.*

Proof. We start with the *special case* in which V has an L-submodule W of codimension one. Since L acts trivially on V/W, by the lemma, we may denote this module F without misleading the reader: $0 \to W \to V \to \mathsf{F} \to 0$ is therefore exact. Using induction on dim W, we can reduce to the case where W is an *irreducible* L-module, as follows. Let W' be a proper nonzero submodule of W. This yields an exact sequence: $0 \to W/W' \to V/W' \to \mathsf{F} \to 0$. By induction, this sequence "splits", i.e., there exists a one dimensional L-submodule of V/W' (say $\widetilde{W}/W'$) complementary to W/W'. So we get another exact sequence: $0 \to W' \to \widetilde{W} \to \mathsf{F} \to 0$. This is like the original situation, except that dim $W' <$ dim W, so induction provides a (one dimensional) submodule X complementary to W' in $\widetilde{W}$: $\widetilde{W} = W' \oplus X$. But $V/W' = W/W' \oplus \widetilde{W}/W'$. It follows that $V = W \oplus X$, since the dimensions add up to dim V and since $W \cap X = 0$.

Now we may assume that W is irreducible. (We may also assume without loss of generality that L acts faithfully on V.) Let $c = c_\phi$ be the Casimir element of ϕ (6.2). Since c commutes with $\phi(L)$, c is actually an *L*-module endomorphism of V; in particular, $c(W) \subset W$ and Ker c is an L-submodule of V. Because L acts trivially on V/W (i.e., $\phi(L)$ sends V into W), c must do likewise (as a linear combination of products of elements $\phi(x)$). So c has trace 0 on V/W. On the other hand, c acts as a scalar on the irreducible L-submodule W (Schur's Lemma); this scalar cannot be 0, because that would force $Tr_V(c) = 0$, contrary to the conclusion of (6.2). It follows that Ker c is a one dimensional L-submodule of V which intersects W trivially. This is the desired complement to W.

Now we can attack the *general case*. Suppose W is any submodule of V: $0 \to W \to V \to V/W \to 0$. Let Hom (V, W) be the space of linear maps $V \to W$, viewed as L-module (6.1). Let $\mathscr{V}$ be the subspace of Hom (V, W) consisting of those maps whose restriction to W is a scalar multiplication. $\mathscr{V}$ is actually an L-submodule: Say $f|_W = a.1_W$; then for $x \in L$, $w \in W$, $(x.f)(w) = x.f(w) - f(x.w) = a(x.w) - a(x.w) = 0$, so $x.f|_W = 0$. Let $\mathscr{W}$ be the subspace of $\mathscr{V}$ consisting of those f whose restriction to W is zero. The preceding calculation shows that $\mathscr{W}$ is also an L-submodule and that L maps $\mathscr{V}$ into $\mathscr{W}$. Moreover, $\mathscr{V}/\mathscr{W}$ has dimension one, because each $f \in \mathscr{V}$ is determined (modulo $\mathscr{W}$) by the scalar $f|_W$. This places us precisely in the situation $0 \to \mathscr{W} \to \mathscr{V} \to \mathsf{F} \to 0$ already treated above.

According to the first part of the proof, $\mathscr{V}$ has a one dimensional submodule complementary to $\mathscr{W}$. Let $f: V \to W$ span it, so after multiplying by a nonzero scalar we may assume that $f|_W = 1_W$. To say that L kills f is just to say that $0 = (x. f)(v) = x.f(v) - f(x.v)$, i.e., that f *is an L-homomorphism*. Therefore Ker f is an L-submodule of V. Since f maps V into W and acts as 1_W on W, we conclude that $V = W \oplus \text{Ker } f$, as desired. $\square$

6.4. Preservation of Jordan decomposition

Weyl's Theorem is of course fundamental for the study of representations of a semisimple Lie algebra L. We offer here a more immediate application, to the problem of showing that the abstract Jordan decomposition (5.4) is compatible with the various linear representations of L.

Theorem. *Let $L \subset \mathfrak{gl}(V)$ be a semisimple linear Lie algebra (V finite dimensional). Then L contains the semisimple and nilpotent parts in $\mathfrak{gl}(V)$ of all its elements. In particular, the abstract and usual Jordan decompositions in L coincide.*

Proof. The last assertion follows from the first, because each type of Jordan decomposition is unique (4.2, 5.4).

Let $x \in L$ be arbitrary, with Jordan decomposition $x = x_s + x_n$ in $\mathfrak{gl}(V)$. The problem is just to show that x_s, x_n lie in L. Since ad $x(L) \subset L$, it follows from Proposition 4.2(c) that ad $x_s(L) \subset L$ and ad $x_n(L) \subset L$, where ad $=$ ad $_{\mathfrak{gl}(V)}$. In other words, x_s, $x_n \in N_{\mathfrak{gl}(V)}(L) = N$, which is a Lie subalgebra of $\mathfrak{gl}(V)$ including L as an ideal. If we could show that $N = L$ we'd be done, but unfortunately this is false: e.g., since $L \subset \mathfrak{sl}(V)$ (Lemma 6.3), the scalars lie in N but not in L. Therefore we need to get x_s, x_n into a smaller subalgebra than N, which can be shown to equal L. If W is any L-submodule of V, define $L_W = \{y \in \mathfrak{gl}(V)|y(W) \subset W$ and $Tr(y|_W) = 0\}$. For example, $L_V = \mathfrak{sl}(V)$. Since $L = [LL]$, it is clear that L lies in all such L_W. Set $L^* =$ intersection of N with all spaces L_W. Clearly, L^* is a subalgebra of N including L as an ideal (but notice that L^* does exclude the scalars). Even more is true: If $x \in L$, then x_s, x_n also lie in L_W, and therefore in L^*.

It remains to prove that $L = L^*$. L^* being a finite dimensional L-module, Weyl's Theorem (6.3) permits us to write $L^* = L + M$ for some L-submodule M, where the sum is direct. But $[L, L^*] \subset L$ (since $L^* \subset N$), so the action of L on M is trivial. Let W be any irreducible L-submodule of V. If $y \in M$, then $[L, y] = 0$, so Schur's Lemma implies that y acts on W as a scalar. On the other hand, $Tr(y|_W) = 0$ because $y \in L_W$. Therefore y acts on W as zero. V can be written as a direct sum of irreducible L-submodules (by Weyl's Theorem), so in fact $y = 0$. This means $M = 0$, $L = L^*$. $\square$

Corollary. *Let L be a semisimple Lie algebra, $\phi: L \to \mathfrak{gl}(V)$ a (finite dimensional) representation of L. If $x = s + n$ is the abstract Jordan decomposition of $x \in L$, then $\phi(x) = \phi(s) + \phi(n)$ is the usual Jordan decomposition of $\phi(x)$.*

Proof. The algebra $\phi(L)$ is spanned by the eigenvectors of $ad_{\phi(L)} \phi(s)$, since L has this property relative to ad s; therefore, $ad_{\phi(L)} \phi(s)$ is semisimple. Similarly, $ad_{\phi(L)} \phi(n)$ is nilpotent, and it commutes with $ad_{\phi(L)} \phi(s)$. Accordingly, $\phi(x) = \phi(s) + \phi(n)$ is the abstract Jordan decomposition of $\phi(x)$ in the semisimple Lie algebra $\phi(L)$ (5.4). Applying the theorem, we get the desired conclusion. $\square$

Exercises

1. Using the standard basis for $L = \mathfrak{sl}(2, \mathsf{F})$, write down the Casimir element of the adjoint representation of L (cf. Exercise 5.5). Do the same thing for the usual (3-dimensional) representation of $\mathfrak{sl}(3, \mathsf{F})$, first computing dual bases relative to the trace form.

2. Let V be an L-module. Prove that V is a direct sum of irreducible submodules if and only if each L-submodule of V possesses a complement.

3. If L is solvable, every irreducible representation of L is one dimensional.

4. Use Weyl's Theorem to give another proof that for L semisimple, ad $L = $ Der L (Theorem 5.3). [If $\delta \in$ Der L, make the direct sum $\mathsf{F} + L$ into an L-module via the rule $x.(a, y) = (0, a\delta(x) + [xy])$. Then consider a complement to the submodule L.]

5. A Lie algebra L for which Rad $L = Z(L)$ is called **reductive**. (Examples: L abelian, L semisimple, $L = \mathfrak{gl}(n, \mathsf{F})$.)

   (a) If L is reductive, then L is a completely reducible ad L-module. [If ad $L \neq 0$, use Weyl's Theorem.] In particular, L is the direct sum of $Z(L)$ and $[LL]$, with $[LL] = 0$ or semisimple.

   (b) If L is a classical linear Lie algebra (1.2), then L is semisimple. [Use part (a), along with Exercises 1.9, 4.1.]

(c) If L is a completely reducible ad L-module, then L is reductive.

(d) If L is reductive, then all finite dimensional representations of L in which $Z(L)$ is represented by semisimple endomorphisms are completely reducible.

6. Let L be a simple Lie algebra. Let $\beta(x, y)$ and $\gamma(x, y)$ be two symmetric associative bilinear forms on L. If β, γ are nondegenerate, prove that β and γ are proportional. [β induces an isomorphism of L onto L^*, and γ in turn induces an isomorphism of L^* onto L. The composite is an L-module isomorphism. But L is an irreducible L-module, so Schur's Lemma applies.]

7. It will be seen later on that $\mathfrak{sl}(n, \mathsf{F})$ is actually *simple*. Assuming this and using Exercise 6, prove that the Killing form κ on $\mathfrak{sl}(n, \mathsf{F})$ is related to the ordinary trace form by $\kappa(x, y) = 2n\, Tr(xy)$.

8. If L is a Lie algebra, then L acts (via ad) on $(L \otimes L)^*$, which may be identified with the space of all bilinear forms β on L. Prove that β is associative if and only if $L.\beta = 0$.

Notes

The proof of Weyl's Theorem is taken from Serre [1]. The original proof was quite different, using integration on compact Lie groups, cf. Freudenthal, de Vries [1]. For Theorem 6.4 we have followed Bourbaki [1].

7. Representations of $\mathfrak{sl}(2, \mathsf{F})$

In this section (as in §6) all modules will be assumed to be finite dimensional over F. L denotes $\mathfrak{sl}(2, \mathsf{F})$, whose standard basis consists of

$$x = \begin{pmatrix} 0 & 1 \\ 0 & 0 \end{pmatrix}, \; y = \begin{pmatrix} 0 & 0 \\ 1 & 0 \end{pmatrix}, \; h = \begin{pmatrix} 1 & 0 \\ 0 & -1 \end{pmatrix}$$

(Example 2.1). Then $[hx] = 2x$, $[hy] = -2y$, $[xy] = h$.

7.1. Weights and maximal vectors

Let V be an arbitrary L-module. Since h is semisimple, Corollary 6.4 implies that h acts diagonally on V. (The assumption that F is algebraically closed insures that all the required eigenvalues already lie in F.) This yields a decomposition of V as direct sum of eigenspaces $V_\lambda = \{v \in V | h.v = \lambda v\}$, $\lambda \in \mathsf{F}$. Of course, the subspace V_λ still makes sense (and is 0) when λ is not an eigenvalue for the endomorphism of V which represents h. Whenever $V_\lambda \neq 0$, we call λ a **weight** of h in V and we call V_λ a **weight space**.

Lemma. *If* $v \in V_\lambda$, *then* $x.v \in V_{\lambda+2}$ *and* $y.v \in V_{\lambda-2}$.

Proof. $h.(x.v) = [h, x].v + x.h.v = 2x.v + \lambda x.v = (\lambda+2)x.v$, and similarly for y. □

Remark. The lemma implies that x, y are represented by nilpotent endomorphisms of V; but this already follows from Theorem 6.4.

Since dim $V < \infty$, and the sum $V = \coprod_{\lambda \in F} V_\lambda$ is direct, there must exist $V_\lambda \neq 0$ such that $V_{\lambda+2} = 0$. (Thanks to the lemma, $x.v = 0$ for any $v \in V_\lambda$.) For such λ, any nonzero vector in V_λ will be called a **maximal vector** of weight λ.

7.2. Classification of irreducible modules

Assume now that V is an irreducible L-module. Choose a maximal vector, say $v_0 \in V_\lambda$; set $v_{-1} = 0$, $v_i = (1/i!)y^i.v_0$ $(i \geq 0)$.

Lemma.   (a) $h.v_i = (\lambda-2i)v_i$,

            (b) $y.v_i = (i+1)v_{i+1}$,

            (c) $x.v_i = (\lambda-i+1)v_{i-1}$ $(i \geq 0)$.

Proof. (a) follows from repeated application of Lemma 7.1, while (b) is just the definition. To prove (c), use induction on i, the case $i = 0$ being clear (since $v_{-1} = 0$, by convention). Observe that

$$ix.v_i = x.y.v_{i-1} \qquad\qquad \text{(by definition)}$$
$$= [x, y].v_{i-1} + y.x.v_{i-1}$$
$$= h.v_{i-1} + y.x.v_{i-1}$$
$$= (\lambda-2(i-1))v_{i-1} + (\lambda-i+2)y.v_{i-2}$$
$$\text{(by (a) and induction)}$$
$$= (\lambda-2i+2)v_{i-1} + (i-1)(\lambda-i+2)v_{i-1} \qquad \text{(by (b))}$$
$$= i(\lambda-i+1)v_{i-1}.$$

Then divide both sides by i. □

Thanks to formula (a), the nonzero v_i are all linearly independent. But dim $V < \infty$. Let m be the smallest integer for which $v_m \neq 0$, $v_{m+1} = 0$; evidently $v_{m+i} = 0$ for all $i > 0$. Taken together, formulas (a)–(c) show that the subspace of V with basis $(v_0, v_1, \ldots, v_m)$ is an L-submodule, different from 0. Because V is irreducible, this subspace must be all of V. Moreover, relative to the ordered basis $(v_0, v_1, \ldots, v_m)$, the matrices of the endomorphisms representing x, y, h can be written down explicitly; notice that h yields a diagonal matrix, while x and y yield (respectively) upper and lower triangular nilpotent matrices.

A closer look at formula (c) reveals a striking fact: for $i = m+1$, the left side is 0, whereas the right side is $(\lambda-m)v_m$. Since $v_m \neq 0$, we conclude that $\lambda = m$. In other words, *the weight of a maximal vector is a nonnegative integer* (one less than dim V). We call it the **highest weight** of V. Moreover, each weight μ occurs with multiplicity one (i.e., dim $V_\mu = 1$ if $V_\mu \neq 0$), by formula (a); in particular, since V determines λ uniquely ($\lambda = \dim V - 1$), the maximal vector v_0 is the only possible one in V (apart from nonzero scalar multiples). To summarize:

Theorem. *Let V be an irreducible module for $L = \mathfrak{sl}(2, \mathsf{F})$.*

(a) *Relative to h, V is the direct sum of weight spaces V_μ, $\mu = m, m-2, \ldots, -(m-2), -m$, where $m + 1 = \dim V$ and $\dim V_\mu = 1$ for each μ.*

(b) *V has (up to nonzero scalar multiples) a unique maximal vector, whose weight (called the highest weight of V) is m.*

(c) *The action of L on V is given explicitly by the above formulas, if the basis is chosen in the prescribed fashion. In particular, there exists at most one irreducible L-module (up to isomorphism) of each possible dimension $m + 1$, $m \geq 0$.* □

Corollary. *Let V be any (finite dimensional) L-module, $L = \mathfrak{sl}(2, \mathsf{F})$. Then the eigenvalues of h on V are all integers, and each occurs along with its negative (an equal number of times). Moreover, in any decomposition of V into direct sum of irreducible submodules, the number of summands is precisely $\dim V_0 + \dim V_1$.*

Proof. If $V = 0$, there is nothing to prove. Otherwise use Weyl's Theorem (6.3) to write V as direct sum of irreducible submodules. The latter are described by the theorem, so the first assertion of the corollary is obvious. For the second, just observe that each irreducible L-module has a unique occurrence of either the weight 0 or else the weight 1 (but not both). □

For the purposes of this chapter, the theorem and corollary just proved are quite adequate. However, it is unreasonable to leave the subject before investigating whether or not $\mathfrak{sl}(2, \mathsf{F})$ does have an irreducible module of each possible highest weight $m = 0, 1, 2, \ldots$ Of course, we already know how to construct suitable modules in low dimensions: the trivial module (dimension 1), the natural representation (dimension 2), the adjoint representation (dimension 3). For arbitrary $m \geq 0$, formulas (a)–(c) of Lemma 7.2 can actually be used to define an irreducible representation of L on an $m + 1$-dimensional vector space over F with basis $(v_0, v_1, \ldots, v_m)$, called $V(m)$. As is customary, the (easy) verification will be left for the reader (Exercise 3). (For a general existence theorem, see (20.3) below.)

One further observation: The symmetry in the structure of $V(m)$ can be made more obvious if we exploit the discussion of *exponentials* in (2.3). Let $\phi: L \to \mathfrak{gl}(V(m))$ be the irreducible representation of highest weight m. Then $\phi(x), \phi(y)$ are nilpotent endomorphisms, in view of the formulas above, so we can define an automorphism of $V(m)$ by $\tau = \exp \phi(x) \exp \phi(-y) \exp \phi(x)$. We may as well assume $m > 0$, so the representation is faithful (L being simple). The discussion in (2.3) shows that conjugating $\phi(h)$ by τ has precisely the same effect as applying $\exp (\mathrm{ad}\ \phi(x)) \exp (\mathrm{ad}\ \phi(-y)) \exp (\mathrm{ad}\ \phi(x))$ to $\phi(h)$. But $\phi(L)$ is isomorphic to L, so this can be calculated just as in (2.3). Conclusion: $\tau\phi(h)\tau^{-1} = -\phi(h)$, or $\tau\phi(h) = -\phi(h)\tau$. From this we see at once that τ sends the basis vector v_i of weight $m - 2i$ to the basis vector v_{m-i} of weight $-(m-2i)$. (The discussion in (2.3) was limited to the special case $m = 1$.) More generally, if V is any finite dimensional L-module, then τ interchanges positive and negative weight spaces.

Exercises

(In these exercises, $L = \mathfrak{sl}(2, \mathsf{F})$.)

1. Use Lie's Theorem to prove the existence of a maximal vector in an arbitrary finite dimensional L-module. [Look at the subalgebra B spanned by h and x.]

2. $M = \mathfrak{sl}(3, \mathsf{F})$ contains a copy of L in its upper left-hand 2×2 position. Write M as direct sum of irreducible L-submodules (M viewed as L-module via the adjoint representation): $V(0) \oplus V(1) \oplus V(1) \oplus V(2)$.

3. Verify that formulas (a)–(c) of Lemma 7.2 do define an irreducible representation of L. [To show that they define a representation, it suffices to show that the matrices corresponding to x, y, h satisfy the same structural equations as x, y, h.]

4. The irreducible representation of L of highest weight m can also be realized "naturally", as follows. Let X, Y be a basis for the two dimensional vector space F^2, on which L acts as usual. Let $\mathscr{R} = \mathsf{F}[X, Y]$ be the polynomial algebra in two variables, and extend the action of L to $\mathscr{R}$ by the derivation rule: $z.fg = (z.f)g + f(z.g)$, for $z \in L, f, g \in \mathscr{R}$. Show that this extension is well defined and that $\mathscr{R}$ becomes an L-module. Then show that the subspace of homogeneous polynomials of degree m, with basis X^m, $X^{m-1}Y, \ldots, XY^{m-1}$, Y^m, is invariant under L and irreducible of highest weight m.

5. Suppose char $\mathsf{F} = p > 0$, $L = \mathfrak{sl}(2, \mathsf{F})$. Prove that the representation $V(m)$ of L constructed as in Exercise 3 or 4 is irreducible so long as the highest weight m is strictly less than p, but reducible when $m = p$.

6. Decompose the tensor product of the two L-modules $V(3)$, $V(7)$ into the sum of irreducible submodules: $V(4) \oplus V(6) \oplus V(8) \oplus V(10)$. Try to develop a general formula for the decomposition of $V(m) \otimes V(n)$.

7. In this exercise we construct certain *infinite dimensional* L-modules. Let $\lambda \in \mathsf{F}$ be an arbitrary scalar. Let $Z(\lambda)$ be a vector space over F with countably infinite basis $(v_0, v_1, v_2, \ldots)$.
   (a) Prove that formulas (a)–(c) of Lemma 7.2 define an L-module structure on $Z(\lambda)$, and that every L-submodule of $Z(\lambda)$ contains at least one maximal vector.
   (b) Suppose $\lambda + 1 = i$ is a nonnegative integer. Prove that v_i is a maximal vector (e.g., $\lambda = -1$, $i = 0$). This induces an L-module homomorphism $Z(\mu) \overset{\phi}{\to} Z(\lambda)$, $\mu = \lambda - 2i$, sending v_0 to v_i. Show that ϕ is a monomorphism, and that Imϕ, $Z(\lambda)/$Imϕ are both irreducible L-modules (but $Z(\lambda)$ fails to be completely reducible when $i > 0$).
   (c) Suppose $\lambda + 1$ is not a nonnegative integer. Prove that $Z(\lambda)$ is irreducible.

ad L = 0

8. Root space decomposition

Throughout this section L denotes a semisimple Lie algebra. We are going to study in detail the structure of L, via its adjoint representation. Our main tools will be the Killing form, and Theorems 6.4, 7.2 (which rely heavily on Weyl's Theorem). The reader should bear in mind the special case $L = \mathfrak{sl}(2, \mathsf{F})$ (or more generally, $\mathfrak{sl}(n, \mathsf{F})$) as a guide to what is going on.

8.1. Maximal toral subalgebras and roots

If L consisted entirely of nilpotent (i.e., ad-nilpotent) elements, then L would be nilpotent (Engel's Theorem). This not being the case, we can find $x \in L$ whose semisimple part x_s in the abstract Jordan decomposition (5.4) is nonzero. This shows that L possesses nonzero subalgebras (e.g., the span of such x_s) consisting of semisimple elements. Call such a subalgebra **toral**. The following lemma is roughly analogous to Engel's Theorem.

Lemma. *A toral subalgebra of L is abelian.*

Proof. Let T be toral. We have to show that $\operatorname{ad}_T x = 0$ for all x in T. Since ad x is diagonalizable (ad x being semisimple and F being algebraically closed), this amounts to showing that $\operatorname{ad}_T x$ has no nonzero eigenvalues. Suppose, on the contrary, that $[xy] = ay$ ($a \neq 0$) for some nonzero y in T. Then $\operatorname{ad}_T y(x) = -ay$ is itself an eigenvector of $\operatorname{ad}_T y$, of eigenvalue 0. On the other hand, we can write x as a linear combination of eigenvectors of $\operatorname{ad}_T y$ (y being semisimple also); after applying $\operatorname{ad}_T y$ to x, all that is left is a combination of eigenvectors which belong to nonzero eigenvalues, if any. This contradicts the preceding conclusion. ⧠

Now fix a **maximal toral subalgebra** H of L, i.e., a toral subalgebra not properly included in any other. (The notation H is less natural than T, but more traditional.) For example, if $L = \mathfrak{sl}(n, \mathsf{F})$, it is easy to verify (Exercise 1) that H can be taken to be the set of diagonal matrices (of trace 0).

Since H is abelian (by the above lemma), $\operatorname{ad}_L H$ is a commuting family of semisimple endomorphisms of L. According to a standard result in linear algebra, $\operatorname{ad}_L H$ is *simultaneously diagonalizable*. In other words, L is the direct sum of the subspaces $L_\alpha = \{x \in L | [hx] = \alpha(h)x \text{ for all } h \in H\}$, where α ranges over H^*. Notice that L_0 is simply $C_L(H)$, the centralizer of H; it includes H, thanks to the lemma. The set of all nonzero $\alpha \in H^*$ for which $L_\alpha \neq 0$ is denoted by Φ; the elements of Φ are called the **roots** of L relative to H (and are finite in number). With this notation we have a **root space decomposition** (or **Cartan decomposition**): $(*) \ L = C_L(H) + \coprod_{\alpha \in \Phi} L_\alpha$. When $L = \mathfrak{sl}(n, \mathsf{F})$, for example, the reader will observe that $(*)$ corresponds to the decomposition of L given by the standard basis (1.2). Our aim in what follows is first to prove that $H = C_L(H)$, then to describe the set of roots in more detail, and ultimately to show that Φ characterizes L completely.

We begin with a few simple observations about the root space decomposition.

Proposition. *For all* $\alpha, \beta \in H^*$, $[L_\alpha L_\beta] \subset L_{\alpha+\beta}$. *If* $x \in L_\alpha$, $\alpha \neq 0$, *then* ad x *is nilpotent. If* $\alpha, \beta \in H^*$, *and* $\alpha + \beta \neq 0$, *then* L_α *is orthogonal to* L_β, *relative to the Killing form* κ *of* L.

Proof. The first assertion follows from the Jacobi identity: $x \in L_\alpha$, $y \in L_\beta$, $h \in H$ imply that ad $h([xy]) = [[hx]y] + [x[hy]] = \alpha(h) [xy] + \beta(h) [xy] = (\alpha+\beta)(h) [xy]$. The second assertion is an immediate consequence of the first.

For the remaining assertion, find $h \in H$ for which $(\alpha+\beta)(h) \neq 0$. Then if $x \in L_\alpha$, $y \in L_\beta$, associativity of the form allows us to write $\kappa([hx], y) = -\kappa([xh], y) = -\kappa(x, [hy])$, or $\alpha(h) \kappa(x, y) = -\beta(h) \kappa(x, y)$, or $(\alpha+\beta)(h) \kappa(x,y) = 0$. This forces $\kappa(x, y) = 0$. □

Corollary. *The restriction of the Killing form to* $L_0 = C_L(H)$ *is nondegenerate.*

Proof. We know from Theorem 5.1 that κ is nondegenerate. On the other hand, L_0 is orthogonal to all L_α ($\alpha \in \Phi$), according to the proposition. If $z \in L_0$ is orthogonal to L_0 as well, then $\kappa(z, L) = 0$, forcing $z = 0$. □

8.2. Centralizer of H

We shall need a fact from linear algebra, whose proof is trivial:

Lemma. *If* x, y *are commuting endomorphisms of a finite dimensional vector space, with* y *nilpotent, then* xy *is nilpotent; in particular,* $Tr(xy) = 0$. □

Proposition. *Let* H *be a maximal toral subalgebra of* L. *Then* $H = C_L(H)$.

Proof. We proceed in steps. Write $C = C_L(H)$.

(1) *C contains the semisimple and nilpotent parts of its elements.* To say that x belongs to $C_L(H)$ is to say that ad x maps the subspace H of L into the subspace 0. By Proposition 4.2, $(\text{ad } x)_s$ and $(\text{ad } x)_n$ have the same property. But by (5.4), $(\text{ad } x)_s = \text{ad } x_s$ and $(\text{ad } x)_n = \text{ad } x_n$.

(2) *All semisimple elements of C lie in H.* If x is semisimple and centralizes H, then $H + Fx$ (which is obviously an abelian subalgebra of L) is toral: the sum of commuting semisimple elements is again semisimple (4.2). By maximality of H, $H + Fx = H$, so $x \in H$.

(3) *The restriction of* κ *to* H *is nondegenerate.* Let $\kappa(h, H) = 0$ for some $h \in H$; we must show that $h = 0$. If $x \in C$ is nilpotent, then the fact that $[xH] = 0$ and the fact that ad x is nilpotent together imply (by the above lemma) that $Tr(\text{ad } x \text{ ad } y) = 0$ for all $y \in H$, or $\kappa(x, H) = 0$. But then (1) and (2) imply that $\kappa(h, C) = 0$, whence $h = 0$ (the restriction of κ to C being nondegenerate by the Corollary to Proposition 8.1).

(4) *C is nilpotent.* If $x \in C$ is semisimple, then $x \in H$ by (2), and $\text{ad}_C x (=0)$ is certainly nilpotent. On the other hand, if $x \in C$ is nilpotent, then $\text{ad}_C x$ is a fortiori nilpotent. Now let $x \in C$ be arbitrary, $x = x_s + x_n$. Since both x_s, x_n

lie in C by (1), $\mathrm{ad}_C\, x$ is the sum of commuting nilpotents and is therefore itself nilpotent. By Engel's Theorem, C is nilpotent.

(5) $H \cap [CC] = 0$. Since κ is associative and $[HC] = 0$, $\kappa(H, [CC]) = 0$. Now use (3).

(6) C is abelian. Otherwise $[CC] \neq 0$. C being nilpotent, by (4), $Z(C) \cap [CC] \neq 0$ (Lemma 3.3). Let $z \neq 0$ lie in this intersection. By (2) and (5), z cannot be semisimple. Its nilpotent part n is therefore nonzero and lies in C, by (1), hence also lies in $Z(C)$ by Proposition 4.2. But then our lemma implies that $\kappa(n, C) = 0$, contrary to Corollary 8.1.

(7) $C = H$. Otherwise C contains a nonzero nilpotent element, x, by (1), (2). According to the lemma and (6), $\kappa(x, y) = Tr(\mathrm{ad}\, x \,\mathrm{ad}\, y) = 0$ for all $y \in C$, contradicting Corollary 8.1. □

Corollary. *The restriction of κ to H is nondegenerate.* □

The corollary allows us to identify H with H^*: to $\phi \in H^*$ corresponds the (unique) element $t_\phi \in H$ satisfying $\phi(h) = \kappa(t_\phi, h)$ for all $h \in H$. In particular, Φ corresponds to the subset $\{t_\alpha; \alpha \in \Phi\}$ of H.

8.3. Orthogonality properties

In this subsection we shall obtain more precise information about the root space decomposition, using the Killing form. We already saw (Proposition 8.1) that $\kappa(L_\alpha, L_\beta) = 0$ if α, $\beta \in H^*$, $\alpha + \beta \neq 0$; in particular, $\kappa(H, L_\alpha) = 0$ for all $\alpha \in \Phi$, so that (Proposition 8.2) the restriction of κ to H is nondegenerate.

Proposition. (a) Φ *spans* H^*.

(b) *If* $\alpha \in \Phi$, *then* $-\alpha \in \Phi$.

(c) *Let* $\alpha \in \Phi$, $x \in L_\alpha$, $y \in L_{-\alpha}$. *Then* $[xy] = \kappa(x, y)t_\alpha$ (t_α *as in* (8.2)).

(d) *If* $\alpha \in \Phi$, *then* $[L_\alpha L_{-\alpha}]$ *is one dimensional, with basis* t_α.

(e) $\alpha(t_\alpha) = \kappa(t_\alpha, t_\alpha) \neq 0$, *for* $\alpha \in \Phi$.

(f) *If* $\alpha \in \Phi$ *and* x_α *is any nonzero element of* L_α, *then there exists* $y_\alpha \in L_{-\alpha}$ *such that* x_α, y_α, $h_\alpha = [x_\alpha y_\alpha]$ *span a three dimensional simple subalgebra of* L *isomorphic to* $\mathfrak{sl}(2, F)$ *via* $x_\alpha \mapsto \begin{pmatrix} 0 & 1 \\ 0 & 0 \end{pmatrix}$, $y_\alpha \mapsto \begin{pmatrix} 0 & 0 \\ 1 & 0 \end{pmatrix}$, $h_\alpha \mapsto \begin{pmatrix} 1 & 0 \\ 0 & -1 \end{pmatrix}$.

(g) $h_\alpha = \dfrac{2t_\alpha}{\kappa(t_\alpha, t_\alpha)}$; $h_\alpha = -h_{-\alpha}$.

Proof. (a) If Φ fails to span H^*, then (by duality) there exists nonzero $h \in H$ such that $\alpha(h) = 0$ for all $\alpha \in \Phi$. But this means that $[h, L_\alpha] = 0$ for all $\alpha \in \Phi$. Since $[hH] = 0$, this in turn forces $[hL] = 0$, or $h \in Z(L) = 0$, which is absurd.

(b) Let $\alpha \in \Phi$. If $-\alpha \notin \Phi$ (i.e., $L_{-\alpha} = 0$), then $\kappa(L_\alpha, L_\beta) = 0$ for all $\beta \in H^*$ (Proposition 8.1). Therefore $\kappa(L_\alpha, L) = 0$, contradicting the nondegeneracy of κ.

(c) Let $\alpha \in \Phi$, $x \in L_\alpha$, $y \in L_{-\alpha}$. Let $h \in H$ be arbitrary. The associativity of κ implies: $\kappa(h, [xy]) = \kappa([hx], y) = \alpha(h)\kappa(x, y) = \kappa(t_\alpha, h)\kappa(x, y) =$

$\kappa(\kappa(x, y)t_\alpha, h) = \kappa(h, \kappa(x, y)t_\alpha)$. This says that H is orthogonal to $[xy] - \kappa(x, y)t_\alpha$, forcing $[xy] = \kappa(x, y)t_\alpha$ (Corollary 8.2).

(d) Part (c) shows that t_α spans $[L_\alpha L_{-\alpha}]$, provided $[L_\alpha L_{-\alpha}] \neq 0$. Let $0 \neq x \in L_\alpha$. If $\kappa(x, L_{-\alpha}) = 0$, then $\kappa(x, L) = 0$ (cf. proof of (b)), which is absurd since κ is nondegenerate. Therefore we can find $0 \neq y \in L_{-\alpha}$ for which $\kappa(x, y) \neq 0$. By (c), $[xy] \neq 0$.

(e) Suppose $\alpha(t_\alpha) = 0$, so that $[t_\alpha x] = 0 = [t_\alpha y]$ for all $x \in L_\alpha$, $y \in L_{-\alpha}$. As in (d), we can find such x, y satisfying $\kappa(x, y) \neq 0$. Modifying one or the other by a scalar, we may as well assume that $\kappa(x, y) = 1$. Then $[xy] = t_\alpha$, by (c). It follows that the subspace S of L spanned by x, y, t_α is a three dimensional solvable algebra, $S \cong \text{ad}_L S \subset \mathfrak{gl}(L)$. In particular, $\text{ad}_L s$ is *nilpotent* for all $s \in [SS]$ (Corollary 4.1A), so $\text{ad}_L t_\alpha$ is both semisimple and nilpotent, i.e., $\text{ad}_L t_\alpha = 0$. This says that $t_\alpha \in Z(L) = 0$, contrary to choice of t_α.

(f) Given $0 \neq x_\alpha \in L_\alpha$, find $y_\alpha \in L_{-\alpha}$ such that $\kappa(x_\alpha, y_\alpha) = \dfrac{2}{\kappa(t_\alpha, t_\alpha)}$. This is possible in view of (e) and the fact that $\kappa(x_\alpha, L_{-\alpha}) \neq 0$. Set $h_\alpha = 2t_\alpha/\kappa(t_\alpha, t_\alpha)$. Then $[x_\alpha y_\alpha] = h_\alpha$, by (c). Moreover, $[h_\alpha x_\alpha] = \dfrac{2}{\alpha(t_\alpha)} [t_\alpha x_\alpha] = \dfrac{2\alpha(t_\alpha)}{\alpha(t_\alpha)} x_\alpha = 2x_\alpha$, and similarly, $[h_\alpha y_\alpha] = -2y_\alpha$. So x_α, y_α, h_α span a three dimensional subalgebra of L with the same multiplication table as $\mathfrak{sl}(2, \mathsf{F})$ (Example 2.1).

(g) Recall that t_α is defined by $\kappa(t_\alpha, h) = \alpha(h)$ ($h \in H$). This shows that $t_\alpha = -t_{-\alpha}$, and in view of the way h_α is defined, the assertion follows. ☐

8.4. Integrality properties

For each pair of roots α, $-\alpha$ (Proposition 8.3(b)), let $S_\alpha \cong \mathfrak{sl}(2, \mathsf{F})$ be a subalgebra of L constructed as in Proposition 8.3(f). Thanks to Weyl's Theorem and Theorem 7.2, we have a complete description of all (finite dimensional) S_α-modules; in particular, we can describe $\text{ad}_L S_\alpha$.

Fix $\alpha \in \Phi$. Consider first the subspace M of L spanned by H along with all root spaces of the form $L_{c\alpha}$ ($c \in \mathsf{F}^*$). This is an S_α-submodule of L, thanks to Proposition 8.1. The weights of h_α on M are the integers 0 and $2c = c\alpha(h_\alpha)$ (for nonzero c such that $L_{c\alpha} \neq 0$), in view of Theorem 7.2. In particular, all c occurring here must be integral multiples of $1/2$. Now S_α acts trivially on $\text{Ker } \alpha$, a subspace of codimension one in H complementary to $\mathsf{F}h_\alpha$, while on the other hand S_α is itself an irreducible S_α-submodule of M. Taken together, $\text{Ker } \alpha$ and S_α exhaust the occurrences of the weight 0 for h_α. Therefore, the only even weights occurring in M are 0, ± 2. This proves that 2α is not a root, i.e., that *twice a root is never a root*. But then $(1/2)\alpha$ cannot be a root either, so 1 cannot occur as a weight of h_α in M. The Corollary of Theorem 7.2 implies that $M = H + S_\alpha$. In particular, $\dim L_\alpha = 1$ (so S_α is uniquely determined as the subalgebra of L generated by L_α and $L_{-\alpha}$), and *the only multiples of a root α which are roots are $\pm \alpha$*.

Next we examine how S_α acts on root spaces L_β, $\beta \neq \pm \alpha$. Set $K = \sum\limits_{i \in \mathsf{Z}}$

$L_{\beta+i\alpha}$. According to the preceding paragraph, each root space is one dimensional and no $\beta+i\alpha$ can equal 0; so K is an S_α-submodule of L, with one dimensional weight spaces for the distinct integral weights $\beta(h_\alpha)+2i$ ($i \in \mathbf{Z}$ such that $\beta+i\alpha \in \Phi$). Obviously, not both 0 and 1 can occur as weights of this form, so the Corollary of Theorem 7.2 implies that K is irreducible. The highest (resp. lowest) weight must be $\beta(h_\alpha)+2q$ (resp. $\beta(h_\alpha)-2r$) if q (resp. r) is the largest integer for which $\beta+q\alpha$ (resp. $\beta-r\alpha$) is a root. Moreover, the weights on K form an arithmetic progression with difference 2 (Theorem 7.2), which implies that the roots $\beta+i\alpha$ form a string (the **α-string through β**) $\beta-r\alpha, \ldots, \beta, \ldots, \beta+q\alpha$. Notice too that $(\beta-r\alpha)(h_\alpha) = -(\beta+q\alpha)(h_\alpha)$, or $\beta(h_\alpha) = r-q$. Finally, observe that if α, β, $\alpha+\beta \in \Phi$, then ad L_α maps L_β onto $L_{\alpha+\beta}$; since these spaces have dimension 1, this says that $[L_\alpha L_\beta] = L_{\alpha+\beta}$.

To summarize:

Proposition. (a) $\alpha \in \Phi$ implies dim $L_\alpha = 1$. In particular, $S_\alpha = L_\alpha + L_{-\alpha} + H_\alpha$ ($H_\alpha = [L_\alpha L_{-\alpha}]$), and for given nonzero $x_\alpha \in L_\alpha$, there exists a unique $y_\alpha \in L_{-\alpha}$ satisfying $[x_\alpha y_\alpha] = h_\alpha$.

(b) If $\alpha \in \Phi$, the only scalar multiples of α which are roots are α and $-\alpha$.

(c) If α, $\beta \in \Phi$, then $\beta(h_\alpha) \in \mathbf{Z}$, and $\beta-\beta(h_\alpha)\alpha \in \Phi$. (The numbers $\beta(h_\alpha)$ are called **Cartan integers**.)

(d) If α, β, $\alpha+\beta \in \Phi$, then $[L_\alpha L_\beta] = L_{\alpha+\beta}$.

(e) Let α, $\beta \in \Phi$, $\beta \neq \pm\alpha$. Let r, q be (respectively) the largest integers for which $\beta-r\alpha$, $\beta+q\alpha$ are roots. Then all $\beta+i\alpha \in \Phi$ ($-r \leq i \leq q$), and $\beta(h_\alpha) = r-q$.

(f) L is generated (as Lie algebra) by the root spaces L_α. □

8.5. Rationality properties. Summary

L is a semisimple Lie algebra (over the algebraically closed field F of characteristic 0), H a maximal toral subalgebra, $\Phi \subset H^*$ the set of roots of L (relative to H), $L = H + \coprod_{\alpha\in\Phi} L_\alpha$ the root space decomposition.

Since the restriction to H of the Killing form is nondegenerate (Corollary 8.2), we may transfer the form to H^*, letting $(\gamma, \delta) = \kappa(t_\gamma, t_\delta)$ for all γ, $\delta \in H^*$. We know that Φ spans H^* (Proposition 8.3(a)), so choose a basis $\alpha_1, \ldots, \alpha_\ell$ of H^* consisting of roots. If $\beta \in \Phi$, we can then write β uniquely as $\beta = \sum_{i=1}^{\ell} c_i\alpha_i$, where $c_i \in \mathsf{F}$. We claim that in fact $c_i \in \mathbf{Q}$. To see this, we use a little linear algebra. For each $j = 1, \ldots, \ell$, $(\beta, \alpha_j) = \sum_{i=1}^{\ell} c_i(\alpha_i, \alpha_j)$, so multiplying both sides by $2/(\alpha_j, \alpha_j)$ yields: $2(\beta, \alpha_j)/(\alpha_j, \alpha_j) = \sum_{i=1}^{\ell} \frac{2(\alpha_i, \alpha_j)}{(\alpha_j, \alpha_j)} c_i$. This may be viewed as a system of ℓ equations in ℓ unknowns c_i, with integral (in particular, rational) coefficients, thanks to Proposition 8.4(c). Since $(\alpha_1, \ldots, \alpha_\ell)$ is a basis of H^*, and the form is nondegenerate, the matrix $((\alpha_i, \alpha_j))_{1 \leq i, j \leq \ell}$ is nonsingular; so the same is true of the coefficient matrix of this system of equations. We conclude that the equations already possess a unique solution over $\mathbf{Q}$, thereby proving our claim.

We have just shown that the $\mathbf{Q}$-subspace $E_\mathbf{Q}$ of H^* spanned by all the roots has $\mathbf{Q}$-dimension $\ell = \dim_F H^*$. Moreover, all inner products of vectors in $E_\mathbf{Q}$ are rational (so we obtain a nondegenerate form on $E_\mathbf{Q}$). Even more is true: Recall that for $\lambda, \mu \in H^*$, $(\lambda, \mu) = \kappa(t_\lambda, t_\mu) = \sum_{\alpha \in \Phi} \alpha(t_\lambda)\alpha(t_\mu) = \sum_{\alpha \in \Phi}$ $(\alpha, \lambda)(\alpha, \mu)$. In particular, $\lambda \in E_\mathbf{Q}$ implies $(\lambda, \lambda) = \sum_{\alpha \in \Phi}(\alpha, \lambda)^2$ is a sum of squares of rational numbers, hence is positive (unless $\lambda = 0$). Therefore, the form on $E_\mathbf{Q}$ is *positive definite*.

Now let E be the real vector space obtained by extending the base field from $\mathbf{Q}$ to $\mathbf{R}$: $E = \mathbf{R} \otimes_\mathbf{Q} E_\mathbf{Q}$. The form extends canonically to E and is positive definite, by the preceding remarks, i.e., E is a euclidean space. Φ contains a basis of E, and $\dim_\mathbf{R} E = \ell$. The following theorem summarizes the basic facts about Φ: cf. Propositions 8.3(a) (b) and 8.4(b) (c).

Theorem. *L, H, Φ, E as above. Then:*

(a) Φ *spans* E, *and* 0 *does not belong to* Φ.

(b) *If* $\alpha \in \Phi$ *then* $-\alpha \in \Phi$, *but no other scalar multiple of* α *is a root.*

(c) *If* $\alpha, \beta \in \Phi$, *then* $\beta - \dfrac{2(\beta, \alpha)}{(\alpha, \alpha)} \alpha \in \Phi$.

(d) *If* $\alpha, \beta \in \Phi$, *then* $\dfrac{2(\beta, \alpha)}{(\alpha, \alpha)} \in \mathbf{Z}$. □

In the language of Chapter III, the theorem asserts that Φ is a **root system** in the real euclidean space E. We have therefore set up a correspondence $(L, H) \mapsto (\Phi, E)$. Pairs (Φ, E) will be completely classified in Chapter III. Later (Chapters IV and V) it will be seen that the correspondence here is actually 1–1, and that the apparent dependence of Φ on the choice of H is not essential.

Exercises

1. If L is a classical linear Lie algebra of type A_ℓ, B_ℓ, C_ℓ, or D_ℓ (see (1.2)), prove that the set of all diagonal matrices in L is a maximal toral subalgebra, of dimension ℓ. (Cf. Exercise 2.8.)

2. For each algebra in Exercise 1, determine the roots and root spaces. How are the various h_α expressed in terms of the basis for H given in (1.2)?

3. If L is of classical type, compute explicitly the restriction of the Killing form to the maximal toral subalgebra described in Exercise 1.

4. If $L = \mathfrak{sl}(2, F)$, prove that each maximal toral subalgebra is one dimensional.

5. If L is semisimple, H a maximal toral subalgebra, prove that H is self-normalizing (i.e., $H = N_L(H)$).

6. Compute the basis of $\mathfrak{sl}(n, \mathsf{F})$ which is dual (via the Killing form) to the standard basis. (Cf. Exercise 5.5.)

7. Let L be semisimple, H a maximal toral subalgebra. If $h \in H$, prove that $C_L(h)$ is *reductive* (in the sense of Exercise 6.5). Prove that H contains elements h for which $C_L(h) = H$; for which h in $\mathfrak{sl}(n, \mathsf{F})$ is this true?

8. For $\mathfrak{sl}(n, \mathsf{F})$ (and other classical algebras), calculate explicitly the root strings and Cartan integers. In particular, prove that all Cartan integers $2(\alpha, \beta)/(\beta, \beta)$, $\alpha \neq \pm\beta$, for $\mathfrak{sl}(n, \mathsf{F})$ are $0, \pm 1$.

9. Prove that every three dimensional semisimple Lie algebra has the same root system as $\mathfrak{sl}(2, \mathsf{F})$, hence is isomorphic to $\mathfrak{sl}(2, \mathsf{F})$.

10. Prove that no four, five or seven dimensional semisimple Lie algebras exist.

11. If $(\alpha, \beta) > 0$, prove that $\alpha - \beta \in \Phi$ $(\alpha, \beta \in \Phi)$. Is the converse true?

Notes

The use of maximal toral subalgebras rather than the more traditional (but equivalent) Cartan subalgebras is suggested by the parallel theory of semisimple algebraic groups: cf. Borel [1], Seligman [2], Winter [1].

Chapter III

Root Systems

9. Axiomatics

9.1. Reflections in a euclidean space

Throughout this chapter we are concerned with a fixed euclidean space E, i.e., a finite dimensional vector space over **R** endowed with a positive definite symmetric bilinear form (α, β). Geometrically, a **reflection** in E is an invertible linear transformation leaving pointwise fixed some **hyperplane** (subspace of codimension one) and sending any vector orthogonal to that hyperplane into its negative. Evidently a reflection is *orthogonal*, i.e., preserves the inner product on E. Any nonzero vector α determines a reflection σ_α, with **reflecting hyperplane** $P_\alpha = \{\beta \in E | (\beta, \alpha) = 0\}$. Of course, nonzero vectors proportional to α yield the same reflection. It is easy to write down an explicit formula:

$\sigma_\alpha(\beta) = \beta - \dfrac{2(\beta, \alpha)}{(\alpha, \alpha)}\alpha$. (This works, because it sends α to $-\alpha$ and fixes all

points in P_α.) Since the number $2(\beta, \alpha)/(\alpha, \alpha)$ occurs frequently, we abbreviate it by $\langle \beta, \alpha \rangle$. Notice that $\langle \beta, \alpha \rangle$ is linear only in the first variable.

For later use we record the following fact.

Lemma. *Let Φ be a finite set which spans E. Suppose all reflections $\sigma_\alpha (\alpha \in \Phi)$ leave Φ invariant. If $\sigma \in GL(E)$ leaves Φ invariant, fixes pointwise a hyperplane P of E, and sends some nonzero $\alpha \in \Phi$ to its negative, then $\sigma = \sigma_\alpha$ (and $P = P_\alpha$).*

Proof. Let $\tau = \sigma\sigma_\alpha$ $(=\sigma\sigma_\alpha^{-1})$. Then $\tau(\Phi) = \Phi$, $\tau(\alpha) = \alpha$, and τ acts as the identity on the subspace $\mathbf{R}\alpha$ as well as on the quotient $E/\mathbf{R}\alpha$. So all eigenvalues of τ are 1, and the minimal polynomial of τ divides $(T-1)^\ell$ ($\ell = $ dim E). On the other hand, since Φ is finite, not all vectors $\beta, \tau(\beta), \ldots, \tau^k(\beta)$ ($\beta \in \Phi$, $k \geq$ Card Φ) can be distinct, so some power of τ fixes β. Choose k large enough so that τ^k fixes all $\beta \in \Phi$. Because Φ spans E, this forces $\tau^k = 1$; so the minimal polynomial of τ divides $T^k - 1$. Combined with the previous step, this shows that τ has minimal polynomial $T-1 = $ g.c.d. $(T^k - 1, (T-1)^\ell)$, i.e., $\tau = 1$. $\square$

9.2. Root systems

A subset Φ of the euclidean space E is called a **root system** in E if the following axioms are satisfied:

(R1) Φ is finite, spans E, and does not contain 0.
(R2) If $\alpha \in \Phi$, the only multiples of α in Φ are $\pm\alpha$.
(R3) If $\alpha \in \Phi$, the reflection σ_α leaves Φ invariant.
(R4) If $\alpha, \beta \in \Phi$, then $\langle \beta, \alpha \rangle \in \mathbf{Z}$.

42

There is some redundancy in the axioms; in particular, both (R2) and (R3) imply that $\Phi = -\Phi$. In the literature (R2) is sometimes omitted, and what we have called a "root system" is then referred to as a "reduced root system" (cf. Exercise 9). Notice that replacement of the given inner product on E by a positive multiple would not affect the axioms, since only *ratios* of inner products occur.

Let Φ be a root system in E. Denote by $\mathscr{W}$ the subgroup of $GL(E)$ generated by the reflections $\sigma_\alpha (\alpha \in \Phi)$. By (R3), $\mathscr{W}$ permutes the set Φ, which by (R1) is finite and spans E. This allows us to identify $\mathscr{W}$ with a subgroup of the symmetric group on Φ; in particular, $\mathscr{W}$ is finite. $\mathscr{W}$ is called the **Weyl group** of Φ, and plays an extremely important role in the sequel. The following lemma shows how certain automorphisms of E act on $\mathscr{W}$ by conjugation.

Lemma. *Let Φ be a root system in E, with Weyl group $\mathscr{W}$. If $\sigma \in GL(E)$ leaves Φ invariant, then $\sigma\sigma_\alpha\sigma^{-1} = \sigma_{\sigma(\alpha)}$ for all $\alpha \in \Phi$, and $\langle \beta, \alpha \rangle = \langle \sigma(\beta), \sigma(\alpha) \rangle$ for all $\alpha, \beta \in \Phi$.*

Proof. $\sigma\sigma_\alpha\sigma^{-1}(\sigma(\beta)) = \sigma\sigma_\alpha(\beta) \in \Phi$, since $\sigma_\alpha(\beta) \in \Phi$. But this equals $\sigma(\beta - \langle \beta, \alpha \rangle \alpha) = \sigma(\beta) - \langle \beta, \alpha \rangle \sigma(\alpha)$. Since $\sigma(\beta)$ runs over Φ as β runs over Φ, we conclude that $\sigma\sigma_\alpha\sigma^{-1}$ leaves Φ invariant, while fixing pointwise the hyperplane $\sigma(P_\alpha)$ and sending $\sigma(\alpha)$ to $-\sigma(\alpha)$. By Lemma 9.1, $\sigma\sigma_\alpha\sigma^{-1} = \sigma_{\sigma(\alpha)}$. But then, comparing the equation above with the equation $\sigma_{\sigma(\alpha)}(\sigma(\beta)) = \sigma(\beta) - \langle \sigma(\beta), \sigma(\alpha) \rangle \sigma(\alpha)$, we also get the second assertion of the lemma. $\square$

There is a natural notion of isomorphism between root systems Φ, Φ' in respective euclidean spaces E, E': Call (Φ, E) and (Φ', E') **isomorphic** if there exists a vector space isomorphism (not necessarily an isometry) ϕ: $E \to E'$ sending Φ onto Φ' such that $\langle \phi(\beta), \phi(\alpha) \rangle = \langle \beta, \alpha \rangle$ for each pair of roots $\alpha, \beta \in \Phi$. It follows at once that $\sigma_{\phi(\alpha)}(\phi(\beta)) = \phi(\sigma_\alpha(\beta))$. Therefore an isomorphism of root systems induces a natural isomorphism $\sigma \mapsto \phi \circ \sigma \circ \phi^{-1}$ of Weyl groups. In view of the lemma above, an automorphism of Φ is the same thing as an automorphism of E leaving Φ invariant. In particular, we can regard $\mathscr{W}$ as a subgroup of Aut Φ (cf. Exercise 6).

It is useful to work not only with α but also with $\alpha^\vee = \dfrac{2\alpha}{(\alpha, \alpha)}$. Call $\Phi^\vee = \{\alpha^\vee | \alpha \in \Phi\}$ the **dual** (or **inverse**) of Φ. It is in fact a root system in E, whose Weyl group is canonically isomorphic to $\mathscr{W}$ (Exercise 2). (In the Lie algebra situation of §8, α corresponds to t_α, while $\alpha^\vee$ corresponds to h_α, under the Killing form identification of H^* with H.)

9.3. Examples

Call $\ell = \dim E$ the **rank** of the root system Φ. When $\ell \leq 2$, we can describe Φ by simply drawing a picture. In view of (R2), there is only one possibility in case $\ell = 1$, labelled (A_1):

$$\xleftarrow{\hspace{1.5cm}} \bullet \xrightarrow{\hspace{1.5cm}}$$
$$-\alpha \qquad\qquad \alpha$$

Of course, this actually is a root system (with Weyl group of order 2); in Lie algebra theory it belongs to $\mathfrak{sl}(2, F)$.

Rank 2 offers more possibilities, four of which are depicted in Figure 1 (these turn out to be the only possibilities). In each case the reader should check the axioms directly and determine $\mathscr{W}$.

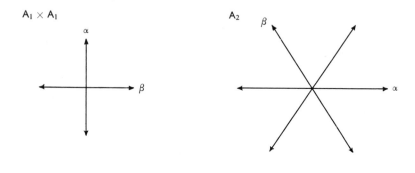

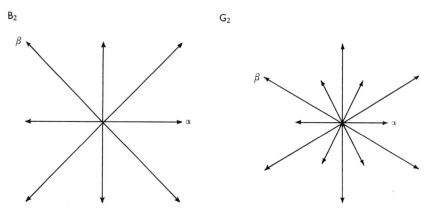

Figure 1

9.4. Pairs of roots

Axiom (R4) limits severely the possible angles occurring between pairs of roots. Recall that the cosine of the angle θ between vectors α, $\beta \in E$ is given by the formula $\|\alpha\| \ \|\beta\| \cos \theta = (\alpha, \beta)$. Therefore, $\langle \beta, \alpha \rangle = \dfrac{2(\beta, \alpha)}{(\alpha, \alpha)} = 2\dfrac{\|\beta\|}{\|\alpha\|} \cos \theta$ and $\langle \alpha, \beta \rangle \langle \beta, \alpha \rangle = 4 \cos^2 \theta$. This last number is a nonnegative integer; but $0 \le \cos^2 \theta \le 1$, and $\langle \alpha, \beta \rangle$, $\langle \beta, \alpha \rangle$ have like sign, so the following possibilities are the only ones when $\alpha \ne \pm \beta$ and $\|\beta\| \ge \|\alpha\|$ (Table 1).

Table 1.

$\langle \alpha, \beta \rangle$	$\langle \beta, \alpha \rangle$	θ	$\|\beta\|^2 / \|\alpha\|^2$
0	0	$\pi/2$	undetermined
1	1	$\pi/3$	1
-1	-1	$2\pi/3$	1
1	2	$\pi/4$	2
-1	-2	$3\pi/4$	2
1	3	$\pi/6$	3
-1	-3	$5\pi/6$	3

The reader will observe that these angles and relative lengths are just the ones portrayed in Figure 1 (9.3). (For $A_1 \times A_1$ it is harmless to change scale in one direction so as to insure that $\|\alpha\| = \|\beta\|$.) The following simple but very useful criterion can be read off from Table 1.

Lemma. *Let* α, β *be nonproportional roots. If* $(\alpha, \beta) > 0$ *(i.e., if the angle between* α *and* β *is strictly acute), then* $\alpha - \beta$ *is a root. If* $(\alpha, \beta) < 0$, *then* $\alpha + \beta$ *is a root.*

Proof. The second assertion follows from the first (applied to $-\beta$ in place of β). Since (α, β) is positive if and only if $\langle \alpha, \beta \rangle$ is, Table 1 shows that one or the other of $\langle \alpha, \beta \rangle$, $\langle \beta, \alpha \rangle$ equals 1. If $\langle \alpha, \beta \rangle = 1$, then $\sigma_\beta(\alpha) = \alpha - \beta \in \Phi$ (R3); similarly, if $\langle \beta, \alpha \rangle = 1$, then $\beta - \alpha \in \Phi$, hence $\sigma_{\beta - \alpha}(\beta - \alpha) = \alpha - \beta \in \Phi$. $\square$

As an application, consider a pair of nonproportional roots α, β. Look at all roots of the form $\beta + i\alpha$ ($i \in \mathbf{Z}$), the **α-string through β**. Let r, $q \in \mathbf{Z}^+$ be the largest integers for which $\beta - r\alpha \in \Phi$, $\beta + q\alpha \in \Phi$ (respectively). If some $\beta + i\alpha \notin \Phi$ ($-r \leq i \leq q$), we can find $p < s$ in this interval such that $\beta + p\alpha \in \Phi$, $\beta + (p+1)\alpha \notin \Phi$, $\beta + (s-1)\alpha \notin \Phi$, $\beta + s\alpha \in \Phi$. But then the lemma implies both $(\alpha, \beta + p\alpha) \geq 0$, $(\alpha, \beta + s\alpha) \leq 0$. Since $p < s$ and $(\alpha, \alpha) > 0$, this is absurd. We conclude that *the* α-*string through* β *is unbroken, from* $\beta - r\alpha$ *to* $\beta + q\alpha$. Now σ_α just adds or subtracts a multiple of α to any root, so this string is invariant under σ_α. Geometrically, it is obvious that σ_α just reverses the string (the reader can easily supply an algebraic proof). In particular, $\sigma_\alpha(\beta + q\alpha) = \beta - r\alpha$. The left side is $\beta - \langle \beta, \alpha \rangle \alpha - q\alpha$, so finally we obtain: $r - q = \langle \beta, \alpha \rangle$ (cf. Proposition 8.1(e)). It follows at once that *root strings are of length at most* 4.

Exercises

(*Unless otherwise specified,* Φ *denotes a root system in* E, *with Weyl group* $\mathscr{W}$.)

1. Let E' be a subspace of E. If a reflection σ_α leaves E' invariant, prove that either $\alpha \in E'$ or else $E' \subset P_\alpha$.

2. Prove that $\Phi^{\vee}$ is a root system in E, whose Weyl group is naturally isomorphic to $\mathscr{W}$; show also that $\langle \alpha^{\vee}, \beta^{\vee} \rangle = \langle \beta, \alpha \rangle$, and draw a picture of $\Phi^{\vee}$ in the cases A_1, A_2, B_2, G_2.

3. In Table 1, show that the order of $\sigma_{\alpha}\sigma_{\beta}$ in $\mathscr{W}$ is (respectively) 2, 3, 4, 6 when $\theta = \pi/2$, $\pi/3$ (or $2\pi/3$), $\pi/4$ (or $3\pi/4$), $\pi/6$ (or $5\pi/6$). [Note that $\sigma_{\alpha}\sigma_{\beta} =$ rotation through 2θ.]

4. Prove that the respective Weyl groups of $A_1 \times A_1$, A_2, B_2, G_2 are dihedral of order 4, 6, 8, 12. If Φ is any root system of rank 2, prove that its Weyl group must be one of these.

5. Show by example that $\alpha - \beta$ may be a root even when $(\alpha, \beta) \leq 0$ (cf. Lemma 9.4).

6. Prove that $\mathscr{W}$ is a normal subgroup of Aut Φ (=group of all isomorphisms of Φ onto itself).

7. Let $\alpha, \beta \in \Phi$ span a subspace E' of E. Prove that $E' \cap \Phi$ is a root system in E'. Prove similarly that $E' \cap (\mathbf{Z}\alpha + \mathbf{Z}\beta)$ is a root system in E' (must this coincide with $E' \cap \Phi$?). More generally, let Φ' be a nonempty subset of Φ such that $\Phi' = -\Phi'$, and such that $\alpha, \beta \in \Phi'$, $\alpha + \beta \in \Phi$ implies $\alpha + \beta \in \Phi'$. Prove that Φ' is a root system in the subspace of E it spans. [Use Table 1].

8. Compute root strings in G_2 to verify the relation $r - q = \langle \beta, \alpha \rangle$.

9. Let Φ be a set of vectors in a euclidean space E, satisfying only (R1), (R3), (R4). Prove that the only possible multiples of $\alpha \in \Phi$ which can be in Φ are $\pm 1/2\, \alpha$, $\pm \alpha$, $\pm 2\alpha$. Verify that $\{\alpha \in \Phi | 2\alpha \notin \Phi\}$ is a root system. *Example*: See Figure 2.

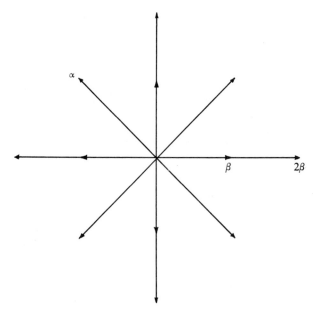

Figure 2

10. Let α, $\beta \in \Phi$. Let the α-string through β be $\beta - r\alpha, \ldots, \beta + q\alpha$, and let the β-string through α be $\alpha - r'\beta, \ldots, \alpha + q'\beta$. Prove that $\dfrac{q(r+1)}{(\beta, \beta)} = \dfrac{q'(r'+1)}{(\alpha, \alpha)}$.

11. Let c be a positive real number. If Φ possesses any roots of squared length c, prove that the set of all such roots is a root system in the subspace of E it spans. Describe the possibilities occurring in Figure 1.

Notes

The axiomatic approach to root systems (as in Serre [2], Bourbaki [2]) has the advantage of providing results which apply simultaneously to Lie algebras, Lie groups, and linear algebraic groups. For historical remarks, consult Bourbaki [2].

10. Simple roots and Weyl group

Throughout this section Φ denotes a root system of rank ℓ in a euclidean space E, with Weyl group $\mathscr{W}$.

10.1. Bases and Weyl chambers

A subset Δ of Φ is called a **base** if:

(B1) Δ is a basis of E,

(B2) each root β can be written as $\beta = \Sigma \, k_\alpha \alpha \; (\alpha \in \Delta)$ with integral coefficients k_α all nonnegative or all nonpositive.

The roots in Δ are then called **simple**. In view of (B1), Card $\Delta = \ell$, and the expression for β in (B2) is unique. This allows us to define the **height** of a root (relative to Δ) by ht$\beta = \sum\limits_{\alpha \in \Delta} k_\alpha$. If all $k_\alpha \geq 0$ (resp. all $k_\alpha \leq 0$), we call β **positive** (resp. **negative**) and write $\beta \succ 0$ (resp. $\beta \prec 0$). The collections of positive and negative roots (relative to Δ) will usually just be denoted Φ^+ and Φ^- (clearly, $\Phi^- = -\Phi^+$). If α and β are positive roots, and $\alpha + \beta$ is a root, then evidently $\alpha + \beta$ is also positive. Actually, Δ defines a *partial order* on E, compatible with the notation $\alpha \succ 0$: define $\beta \prec \alpha$ iff $\alpha - \beta$ is a sum of positive roots (equivalently, of simple roots) or $\beta = \alpha$.

The only problem with the definition of base is that it fails to guarantee *existence*. In the examples shown in (9.3), the roots labelled α, β do form a base in each case (verify!). Notice there that the angle between α and β is *obtuse*, i.e., $(\alpha, \beta) \leq 0$. This is no accident.

Lemma. *If Δ is a base of Φ, then $(\alpha, \beta) \leq 0$ for $\alpha \neq \beta$ in Δ, and $\alpha - \beta$ is not a root.*

Proof. Otherwise $(\alpha, \beta) > 0$. Since $\alpha \neq \beta$, by assumption, and since obviously $\alpha \neq -\beta$, Lemma 9.4 then says that $\alpha - \beta$ is a root. But this violates (B2). $\square$

Our goal is the proof of the following theorem.

Theorem. Φ *has a base.*

The proof will in fact yield a concrete method for constructing all possible bases. For each vector $\gamma \in E$, define $\Phi^+(\gamma) = \{\alpha \in \Phi | (\gamma, \alpha) > 0\} =$ the set of roots lying on the "positive" side of the hyperplane orthogonal to γ. It is an elementary fact in euclidean geometry that the union of the finitely many hyperplanes P_α ($\alpha \in \Phi$) cannot exhaust E (we leave to the reader the task of formulating a rigorous proof). Call $\gamma \in E$ **regular** if $\gamma \in E - \bigcup_{\alpha \in \Phi} P_\alpha$, **singular** otherwise. When γ is regular, it is clear that $\Phi = \Phi^+(\gamma) \cup -\Phi^+(\gamma)$. This is the case we shall now pursue. Call $\alpha \in \Phi^+(\gamma)$ **decomposable** if $\alpha = \beta_1 + \beta_2$ for some $\beta_i \in \Phi^+(\gamma)$, **indecomposable** otherwise. Now it suffices to prove the following statement.

Theorem'. *Let* $\gamma \in E$ *be regular. Then the set* $\Delta(\gamma)$ *of all indecomposable roots in* $\Phi^+(\gamma)$ *is a base of* Φ, *and every base is obtainable in this manner.*

Proof. This will proceed in steps.

(1) *Each root in* $\Phi^+(\gamma)$ *is a nonnegative* **Z**-*linear combination of* $\Delta(\gamma)$. Otherwise some $\alpha \in \Phi^+(\gamma)$ cannot be so written; choose α so that (γ, α) is as small as possible. Obviously α itself cannot be in $\Delta(\gamma)$, so $\alpha = \beta_1 + \beta_2$ ($\beta_i \in \Phi^+(\gamma)$), whence $(\gamma, \alpha) = (\gamma, \beta_1) + (\gamma, \beta_2)$. But each of the (γ, β_i) is positive, so β_1 and β_2 must each be a nonnegative **Z**-linear combination of $\Delta(\gamma)$ (to avoid contradicting the minimality of (γ, α)), whence α also is. This contradiction proves the original assertion.

(2) *If* $\alpha, \beta \in \Delta(\gamma)$, *then* $(\alpha, \beta) \le 0$ *unless* $\alpha = \beta$. Otherwise $\alpha - \beta$ is a root (Lemma 9.4), since β clearly cannot be $-\alpha$, so $\alpha - \beta$ or $\beta - \alpha$ is in $\Phi^+(\gamma)$. In the first case, $\alpha = \beta + (\alpha - \beta)$, which says that α is decomposable; in the second case, $\beta = \alpha + (\beta - \alpha)$ is decomposable. This contradicts the assumption.

(3) $\Delta(\gamma)$ *is a linearly independent set.* Suppose $\Sigma r_\alpha \alpha = 0$ ($\alpha \in \Delta(\gamma)$, $r_\alpha \in \mathbf{R}$). Separating the indices α for which $r_\alpha > 0$ from those for which $r_\alpha < 0$, we can rewrite this as $\Sigma s_\alpha \alpha = \Sigma t_\beta \beta$ ($s_\alpha, t_\beta > 0$, the sets of α's and β's being disjoint). Call $\varepsilon = \Sigma s_\alpha \alpha$. Then $(\varepsilon, \varepsilon) = \sum_{\alpha, \beta} s_\alpha t_\beta (\alpha, \beta) \le 0$ by step (2), forcing $\varepsilon = 0$. Then $0 = (\gamma, \varepsilon) = \Sigma s_\alpha (\gamma, \alpha)$, forcing all $s_\alpha = 0$. Similarly, all $t_\beta = 0$. (This argument actually shows that *any set of vectors lying strictly on one side of a hyperplane in* E *and forming pairwise obtuse angles must be linearly independent.*)

(4) $\Delta(\gamma)$ *is a base of* Φ. Since $\Phi = \Phi^+(\gamma) \cup - \Phi^+(\gamma)$, the requirement (B2) is satisfied thanks to step (1). It also follows that $\Delta(\gamma)$ spans E, which combined with step (3) yields (B1).

(5) *Each base* Δ *of* Φ *has the form* $\Delta(\gamma)$ *for some regular* $\gamma \in E$. Given Δ, select $\gamma \in E$ so that $(\gamma, \alpha) > 0$ for all $\alpha \in \Delta$. (This is possible, because the intersection of "positive" open half-spaces associated with any basis of E is nonvoid (Exercise 7).) In view of (B2), γ is regular and $\Phi^+ \subset \Phi^+(\gamma)$, $\Phi^- \subset -\Phi^+(\gamma)$ (so equality must hold in each instance). Since $\Phi^+ = \Phi^+(\gamma)$,

Δ clearly consists of indecomposable elements, i.e., $\Delta \subset \Delta(\gamma)$. But Card Δ = Card $\Delta(\gamma)$ = ℓ, so $\Delta = \Delta(\gamma)$. $\square$

It is useful to introduce a bit of terminology. The hyperplanes P_α $(\alpha \in \Phi)$ partition E into finitely many regions; the connected components of $\mathsf{E} - \bigcup_\alpha P_\alpha$ are called the (open) **Weyl chambers** of E. Each regular $\gamma \in \mathsf{E}$ therefore belongs to precisely one Weyl chamber, denoted $\mathfrak{C}(\gamma)$. To say that $\mathfrak{C}(\gamma) = \mathfrak{C}(\gamma')$ is just to say that γ, γ' lie on the same side of each hyperplane P_α $(\alpha \in \Phi)$, i.e., that $\Phi^+(\gamma) = \Phi^+(\gamma')$, or $\Delta(\gamma) = \Delta(\gamma')$. This shows that *Weyl chambers are in natural 1-1 correspondence with bases.* Write $\mathfrak{C}(\Delta) = \mathfrak{C}(\gamma)$ if $\Delta = \Delta(\gamma)$, and call this the **fundamental Weyl chamber relative to Δ**. $\mathfrak{C}(\Delta)$ is the open convex set (intersection of open half-spaces) consisting of all $\gamma \in \mathsf{E}$ which satisfy the inequalities $(\gamma, \alpha) > 0$ $(\alpha \in \Delta)$. In rank 2, it is easy to draw the appropriate picture; this is done in Figure 1 for type A_2. Here there are six chambers, the shaded one being fundamental relative to the base $\{\alpha, \beta\}$.

The Weyl group obviously sends one Weyl chamber onto another: explicitly, $\sigma(\mathfrak{C}(\gamma)) = \mathfrak{C}(\sigma\gamma)$, if $\sigma \in \mathscr{W}$ and γ is regular. On the other hand, $\mathscr{W}$ permutes bases: σ sends Δ to $\sigma(\Delta)$, which is again a base (why?). These two actions of $\mathscr{W}$ are in fact compatible with the above correspondence

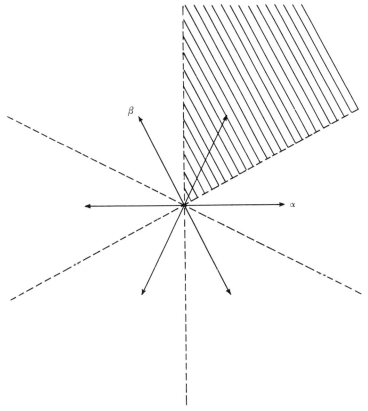

Figure 1

between Weyl chambers and bases; we have $\sigma(\Delta(\gamma)) = \Delta(\sigma\gamma)$, because $(\sigma\gamma, \sigma\alpha) = (\gamma, \alpha)$.

10.2. Lemmas on simple roots

Let Δ be a fixed base of Φ. We prove here several very useful lemmas about the behavior of simple roots.

Lemma A. *If α is positive but not simple, then $\alpha - \beta$ is a root (necessarily positive) for some $\beta \in \Delta$.*

Proof. If $(\alpha, \beta) \le 0$ for all $\beta \in \Delta$, the parenthetic remark in step (3) in (10.1) would apply, showing that $\Delta \cup \{\alpha\}$ is a linearly independent set. This is absurd, since Δ is already a basis of E. So $(\alpha, \beta) > 0$ for some $\beta \in \Delta$ and then $\alpha - \beta \in \Phi$ (Lemma 9.4, which applies since β cannot be proportional to α). Write $\alpha = \sum_{\gamma \in \Delta} k_\gamma \gamma$ (all $k_\gamma \ge 0$, some $k_\gamma > 0$ for $\gamma \ne \beta$). Subtracting β from α yields a $\mathbf{Z}$-linear combination of simple roots with at least one positive coefficient. This forces all coefficients to be positive, thanks to the uniqueness of expression in (B2). $\square$

Corollary. *Each $\beta \in \Phi^+$ can be written in the form $\alpha_1 + \ldots + \alpha_k$ ($\alpha_i \in \Delta$, not necessarily distinct) in such a way that each partial sum $\alpha_1 + \ldots + \alpha_i$ is a root.*

Proof. Use the lemma and induction on $\mathrm{ht}\beta$. $\square$

Lemma B. *Let α be simple. Then σ_α permutes the positive roots other than α.*

Proof. Let $\beta \in \Phi^+ - \{\alpha\}$, $\beta = \sum_{\gamma \in \Delta} k_\gamma \gamma$ ($k_\gamma \in \mathbf{Z}^+$). It is clear that β is not proportional to α. Therefore, $k_\gamma \ne 0$ for some $\gamma \ne \alpha$. But the coefficient of γ in $\sigma_\alpha(\beta) = \beta - \langle \beta, \alpha \rangle \alpha$ is still k_γ. In other words, $\sigma_\alpha(\beta)$ has at least one positive coefficient (relative to Δ), forcing it to be positive. Moreover, $\sigma_\alpha(\beta) \ne \alpha$, since α is the image of $-\alpha$. $\square$

Corollary. *Set $\delta = \frac{1}{2} \sum_{\beta > 0} \beta$. Then $\sigma_\alpha(\delta) = \delta - \alpha$ for all $\alpha \in \Delta$.*

Proof. Obvious from the lemma. $\square$

Lemma C. *Let $\alpha_1, \ldots, \alpha_t \in \Delta$ (not necessarily distinct). Write $\sigma_i = \sigma_{\alpha_i}$. If $\sigma_1 \ldots \sigma_{t-1}(\alpha_t)$ is negative, then for some index $1 \le s < t$, $\sigma_1 \ldots \sigma_t = \sigma_1 \ldots \sigma_{s-1} \sigma_{s+1} \ldots \sigma_{t-1}$.*

Proof. Write $\beta_i = \sigma_{i+1} \ldots \sigma_{t-1}(\alpha_t)$, $0 \le i \le t-2$, $\beta_{t-1} = \alpha_t$. Since $\beta_0 \prec 0$ and $\beta_{t-1} \succ 0$, we can find a smallest index s for which $\beta_s \succ 0$. Then $\sigma_s(\beta_s) = \beta_{s-1} \prec 0$, and Lemma B forces $\beta_s = \alpha_s$. In general (Lemma 9.2), $\sigma \in \mathscr{W}$ implies $\sigma_{\sigma(\alpha)} = \sigma \sigma_\alpha \sigma^{-1}$; so in particular, $\sigma_s = (\sigma_{s+1} \ldots \sigma_{t-1})\sigma_t (\sigma_{t-1} \ldots \sigma_{s+1})$ which yields the lemma. $\square$

Corollary. *If $\sigma = \sigma_1 \ldots \sigma_t$ is an expression for $\sigma \in \mathscr{W}$ in terms of reflections corresponding to simple roots, with t as small as possible, then $\sigma(\alpha_t) \prec 0$.* $\square$

10.3. The Weyl group

Now we are in a position to prove that $\mathscr{W}$ permutes the bases of Φ (or, equivalently, the Weyl chambers) in a simply transitive fashion and that $\mathscr{W}$ is generated by the "simple reflections" relative to any base Δ (i.e., by the σ_α for $\alpha \in \Delta$).

Theorem. *Let Δ be a base of Φ.*

(a) *If $\gamma \in E$, γ regular, there exists $\sigma \in \mathscr{W}$ such that $(\sigma(\gamma), \alpha) > 0$ for all $\alpha \in \Delta$ (so $\mathscr{W}$ acts transitively on Weyl chambers).*

(b) *If Δ' is another base of Φ, then $\sigma(\Delta') = \Delta$ for some $\sigma \in \mathscr{W}$ (so $\mathscr{W}$ acts transitively on bases).*

(c) *If α is any root, there exists $\sigma \in \mathscr{W}$ such that $\sigma(\alpha) \in \Delta$.*

(d) *$\mathscr{W}$ is generated by the σ_α $(\alpha \in \Delta)$.*

(e) *If $\sigma(\Delta) = \Delta$, $\sigma \in \mathscr{W}$, then $\sigma = 1$ (so $\mathscr{W}$ acts simply transitively on bases).*

Proof. Let $\mathscr{W}'$ be the subgroup of $\mathscr{W}$ generated by all σ_α $(\alpha \in \Delta)$. We shall prove (a)–(c) for $\mathscr{W}'$, then deduce that $\mathscr{W}' = \mathscr{W}$.

(a) Write $\delta = \frac{1}{2} \sum_{\alpha > 0} \alpha$, and choose $\sigma \in \mathscr{W}'$ for which $(\sigma(\gamma), \delta)$ is as big as possible. If α is simple, then of course $\sigma_\alpha \sigma$ is also in $\mathscr{W}'$, so the choice of σ implies that $(\sigma(\gamma), \delta) \geq (\sigma_\alpha \sigma(\gamma), \delta) = (\sigma(\gamma), \sigma_\alpha(\delta)) = (\sigma(\gamma), \delta - \alpha) = (\sigma(\gamma), \delta) - (\sigma(\gamma), \alpha)$ (Corollary to Lemma 10.2B). This forces $(\sigma(\gamma), \alpha) \geq 0$ for all $\alpha \in \Delta$. Since γ is regular, we cannot have $(\sigma(\gamma), \alpha) = 0$ for any α, because then γ would be orthogonal to $\sigma^{-1}\alpha$. So all the inequalities are strict. Therefore $\sigma(\gamma)$ lies in the fundamental Weyl chamber $\mathfrak{C}(\Delta)$, and σ sends $\mathfrak{C}(\gamma)$ to $\mathfrak{C}(\Delta)$ as desired.

(b) Since $\mathscr{W}'$ permutes the Weyl chambers, by (a), it also permutes the bases of Φ (transitively).

(c) In view of (b), it suffices to prove that each root belongs to at least one base. Since the only roots proportional to α are $\pm \alpha$, the hyperplanes P_β $(\beta \neq \pm \alpha)$ are distinct from P_α, so there exists $\gamma \in P_\alpha$, $\gamma \notin P_\beta$ (all $\beta \neq \pm \alpha$) (why?). Choose γ' close enough to γ so that $(\gamma', \alpha) = \varepsilon > 0$ while $|(\gamma', \beta)| > \varepsilon$ for all $\beta \neq \pm \alpha$. Evidently α then belongs to the base $\Delta(\gamma')$.

(d) To prove $\mathscr{W}' = \mathscr{W}$, it is enough to show that each reflection σ_α $(\alpha \subset \Phi)$ is in $\mathscr{W}'$. Using (c), find $\sigma \in \mathscr{W}'$ such that $\beta = \sigma(\alpha) \in \Delta$. Then $\sigma_\beta = \sigma_{\sigma(\alpha)} = \sigma\sigma_\alpha\sigma^{-1}$, so $\sigma_\alpha = \sigma^{-1}\sigma_\beta\sigma \in \mathscr{W}'$.

(e) Let $\sigma(\Delta) = \Delta$, but $\sigma \neq 1$. If σ is written minimally as a product of one or more simple reflections (which is possible, thanks to (d)), then the Corollary to Lemma 10.2C is contradicted. $\square$

We can use the lemmas of (10.2) to explore more precisely the significance of the generation of $\mathscr{W}$ by simple reflections.

When $\sigma \in \mathscr{W}$ is written as $\sigma_{\alpha_1} \ldots \sigma_{\alpha_t}$ $(\alpha_i \in \Delta$, t minimal), we call the expression **reduced**, and write $\ell(\sigma) = t$: this is the **length** of σ, relative to Δ. By definition, $\ell(1) = 0$. We can characterize length in another way, as follows. Define $n(\sigma) =$ number of positive roots α for which $\sigma(\alpha) \prec 0$.

Lemma A. *For all* $\sigma \in \mathcal{W}$, $\ell(\sigma) = n(\sigma)$.

Proof. Proceed by induction on $\ell(\sigma)$. The case $\ell(\sigma) = 0$ is clear: $\ell(\sigma) = 0$ implies $\sigma = 1$, so $n(\sigma) = 0$. Assume the lemma true for all $\tau \in \mathcal{W}$ with $\ell(\tau) < \ell(\sigma)$. Write σ in reduced form as $\sigma = \sigma_{\alpha_1} \ldots \sigma_{\alpha_t}$, and set $\alpha = \alpha_t$. By the Corollary of Lemma 10.2C, $\sigma(\alpha) \prec 0$. Then Lemma 10.2B implies that $n(\sigma\sigma_\alpha) = n(\sigma) - 1$. On the other hand, $\ell(\sigma\sigma_\alpha) = \ell(\sigma) - 1 < \ell(\sigma)$, so by induction $\ell(\sigma\sigma_\alpha) = n(\sigma\sigma_\alpha)$. Combining these statements, we get $\ell(\sigma) = n(\sigma)$. $\quad\square$

Next we look more closely at the simply transitive action of $\mathcal{W}$ on Weyl chambers (parts (a) and (e) of the theorem). The next lemma shows that the closure $\overline{\mathfrak{C}(\Delta)}$ of the fundamental Weyl chamber relative to Δ is a **fundamental domain** for the action of $\mathcal{W}$ on E, i.e., each vector in E is $\mathcal{W}$-conjugate to precisely one point of this set.

Lemma B. *Let* $\gamma \in \overline{\mathfrak{C}(\Delta)}$. *Then* $\sigma\gamma \prec \gamma$ *for all* $\sigma \in \mathcal{W}$. *If* $\gamma \in \mathfrak{C}(\Delta)$, *then* $\sigma\gamma = \gamma$ *only when* $\sigma = 1$.

Proof. We may assume $\sigma \neq 1$. Write $\sigma = \sigma_{i(1)} \ldots \sigma_{i(t)}$ in reduced form as a product of simple reflections, where $\sigma_i = \sigma_{\alpha_i}$, $\Delta = \{\alpha_1, \ldots, \alpha_l\}$. Obviously each partial product $\sigma_{i(t)} \ldots \sigma_{i(s)}$ ($1 \leq s \leq t$) in σ^{-1} is also in reduced form, so the Corollary of Lemma 10.2C implies that $\sigma_{i(t)} \ldots \sigma_{i(s+1)}(\alpha_{i(s)})$ is positive. The inner product being invariant under $\mathcal{W}$, we can write $(\sigma_{i(s+1)} \ldots \sigma_{i(t)}\gamma, \alpha_{i(s)}) = (\gamma, \sigma_{i(t)} \ldots \sigma_{i(s+1)}\alpha_{i(s)})$, which is nonnegative because of the assumption on γ. In other words, $\sigma\gamma$ is obtained from γ by subtracting $\alpha_{i(t)}, \alpha_{i(t-1)}, \ldots, \alpha_{i(1)}$ the appropriate number of times; therefore, $\sigma\gamma \prec \gamma$. The argument shows also that if $\sigma\gamma = \gamma$, then $\sigma_{i(t)}\gamma = \gamma$, or $(\gamma, \alpha_{i(t)}) = 0$, whence γ cannot lie in $\mathfrak{C}(\Delta)$. $\quad\square$

10.4. Irreducible root systems

Φ is called **irreducible** if it cannot be partitioned into the union of two proper subsets such that each root in one set is orthogonal to each root in the other. (In (9.3), A_1, A_2, B_2, G_2 are irreducible, while $A_1 \times A_1$ is not.) Suppose Δ is a base of Φ. We claim that Φ *is irreducible if and only if* Δ *cannot be partitioned in the way just stated.* In one direction, let $\Phi = \Phi_1 \cup \Phi_2$, with $(\Phi_1, \Phi_2) = 0$. Unless Δ is wholly contained in Φ_1 or Φ_2, this induces a similar partition of Δ; but $\Delta \subset \Phi_1$ implies $(\Delta, \Phi_2) = 0$, or $(E, \Phi_2) = 0$, since Δ spans E. This shows that the "if" holds. Conversely, let Φ be irreducible, but $\Delta = \Delta_1 \cup \Delta_2$ with $(\Delta_1, \Delta_2) = 0$. Each root is conjugate to a simple root (Theorem 10.3(c)), so $\Phi = \Phi_1 \cup \Phi_2$, Φ_i the set of roots having a conjugate in Δ_i. Recall that $(\alpha, \beta) = 0$ implies $\sigma_\alpha\sigma_\beta = \sigma_\beta\sigma_\alpha$. Since $\mathcal{W}$ is generated by the σ_α ($\alpha \in \Delta$), the formula for a reflection makes it clear that each root in Φ_i is gotten from one in Δ_i by adding or subtracting elements of Δ_i. Therefore, Φ_i lies in the subspace E_i of E spanned by Δ_i, and we see that $(\Phi_1, \Phi_2) = 0$. This forces $\Phi_1 = \varnothing$ or $\Phi_2 = \varnothing$, whence $\Delta_1 = \varnothing$ or $\Delta_2 = \varnothing$.

Lemma A. *Let* Φ *be irreducible. Relative to the partial ordering* $\prec$, *there* *is a unique maximal root β (in particular, $\alpha \neq \beta$ implies $ht\alpha < ht\beta$, and (β, α) ≥ 0 for all $\alpha \in \Delta$). If $\beta = \Sigma k_\alpha \alpha$ ($\alpha \in \Delta$) then all $k_\alpha > 0$.*

Proof. Let $\beta = \Sigma k_\alpha \alpha$ ($\alpha \in \Delta$) be maximal in the ordering; evidently $\beta \succ 0$. If $\Delta_1 = \{\alpha \in \Delta | k_\alpha > 0\}$ and $\Delta_2 = \{\alpha \in \Delta | k_\alpha = 0\}$, then $\Delta = \Delta_1 \cup \Delta_2$ is a partition. Suppose Δ_2 is nonvoid. Then $(\alpha, \beta) \leq 0$ for any $\alpha \in \Delta_2$ (Lemma 10.1); since Φ is irreducible, at least one $\alpha \in \Delta_2$ must be nonorthogonal to Δ_1, forcing $(\alpha, \alpha') < 0$ for some $\alpha' \in \Delta_1$, whence $(\alpha, \beta) < 0$. This implies that $\beta + \alpha$ is a root (Lemma 9.4), contradicting the maximality of β. Therefore Δ_2 is empty and all $k_\alpha > 0$. This argument shows also that $(\alpha, \beta) \geq 0$ for all $\alpha \in \Delta$ (with $(\alpha, \beta) > 0$ for at least one α, since Δ spans E). Now let β' be another maximal root. The preceding argument applies to β', so β' involves (with positive coefficient) at least one $\alpha \in \Delta$ for which $(\alpha, \beta) > 0$. It follows that $(\beta', \beta) > 0$, and $\beta - \beta'$ is a root (Lemma 9.4) unless $\beta = \beta'$. But if $\beta - \beta'$ is a root, then either $\beta \prec \beta'$ or else $\beta' \prec \beta$, which is absurd. So β is unique. $\square$

Lemma B. *Let Φ be irreducible. Then $\mathscr{W}$ acts irreducibly on E. In particular, the $\mathscr{W}$-orbit of a root α spans E.*

Proof. The span of the $\mathscr{W}$-orbit of a root is a (nonzero) $\mathscr{W}$-invariant subspace of E, so the second statement follows from the first. As to the first, let E' be a nonzero subspace of E invariant under $\mathscr{W}$. The orthogonal complement E'' of E' is also $\mathscr{W}$-invariant, and $E = E' \oplus E''$. It is trivial to verify that for $\alpha \in \Phi$, either $\alpha \in E'$ or else $E' \subset P_\alpha$, since $\sigma_\alpha(E') = E'$ (Exercise 9.1). Thus, $\alpha \notin E'$ implies $\alpha \in E''$, so each root lies in one subspace or the other. This partitions Φ into orthogonal subsets, forcing one or the other to be empty. Since Φ spans E, we conclude that $E' = E$. $\square$

Lemma C. *Let Φ be irreducible. Then at most two root lengths occur in Φ, and all roots of a given length are conjugate under $\mathscr{W}$.*

Proof. If α, β are arbitrary roots, then not all $\sigma(\alpha)$ ($\sigma \in \mathscr{W}$) can be orthogonal to β, since the $\sigma(\alpha)$ span E (Lemma B). If $(\alpha, \beta) \neq 0$, we know (cf. (9.4)) that the possible ratios of squared root lengths of α, β are 1, 2, 3, $1/2$, $1/3$. These two remarks easily imply the first assertion of the lemma, since the presence of three root lengths would yield also a ratio $3/2$. Now let α, β have equal length. After replacing one of these by a $\mathscr{W}$-conjugate (as above), we may assume them to be nonorthogonal (and distinct: otherwise we're done!). According to (9.4), this in turn forces $\langle \alpha, \beta \rangle = \langle \beta, \alpha \rangle = \pm 1$. Replacing β (if need be) by $-\beta = \sigma_\beta(\beta)$, we may assume that $\langle \alpha, \beta \rangle = 1$. Therefore, $(\sigma_\alpha \sigma_\beta \sigma_\alpha)(\beta) = \sigma_\alpha \sigma_\beta(\beta - \alpha) = \sigma_\alpha(-\beta - \alpha + \beta) = \alpha$. $\square$

In case Φ is irreducible, with two distinct root lengths, we speak of **long** and **short roots**. (If all roots are of equal length, it is conventional to call all of them long.)

Lemma D. *Let Φ be irreducible, with two distinct root lengths. Then the maximal root β of Lemma A is long.*

Proof. Let $\alpha \in \Phi$ be arbitrary. It will suffice to show that $(\beta, \beta) \geq (\alpha, \alpha)$. For this we may replace α by a $\mathscr{W}$-conjugate lying in the closure of the fundamental Weyl chamber (relative to Δ). Since $\beta - \alpha \succ 0$ (Lemma A), we have $(\gamma, \beta - \alpha) \geq 0$ for any $\gamma \in \overline{\mathfrak{C}(\Delta)}$. This fact, applied to the cases $\gamma = \beta$ (cf. Lemma A) and $\gamma = \alpha$, yields $(\beta, \beta) \geq (\beta, \alpha) \geq (\alpha, \alpha)$. $\square$

Exercises

1. Let $\Phi^\vee$ be the dual system of Φ, $\Delta^\vee = \{\alpha^\vee | \alpha \in \Delta\}$. Prove that $\Delta^\vee$ is a base of $\Phi^\vee$. [Look at the effect of a simple reflection on an element of $\Delta^\vee$, and use Theorem 10.3.]

2. If Δ is a base of Φ, prove that the set $(\mathbf{Z}\alpha + \mathbf{Z}\beta) \cap \Phi$ ($\alpha \neq \beta$ in Δ) is a root system of rank 2 in the subspace of E spanned by α, β (cf. Exercise 9.7). Generalize to an arbitrary subset of Δ.

3. Prove that each root system of rank 2 is isomorphic to one of those listed in (9.3).

4. Verify the Corollary of Lemma 10.2A directly for G_2.

5. If $\sigma \in \mathscr{W}$ can be written as a product of t simple reflections, prove that t has the same parity as $\ell(\sigma)$.

6. Define a function $sn: \mathscr{W} \to \{\pm 1\}$ by $sn(\sigma) = (-1)^{\ell(\sigma)}$. Prove that sn is a homomorphism (cf. the case A_2, where $\mathscr{W}$ is isomorphic to the symmetric group $\mathscr{S}_3$).

7. Prove that the intersection of "positive" open half-spaces associated with any basis $\gamma_1, \ldots, \gamma_\ell$ of E is nonvoid. [Let δ_i be the projection of γ_i on the subspace spanned by all basis vectors except γ_i. Then $\gamma = \Sigma r_i \delta_i$ is in the required intersection provided all $r_i > 0$.]

8. Let Δ be a base of Φ, $\alpha \neq \beta$ simple roots, $\Phi_{\alpha\beta}$ the rank 2 root system in $\mathsf{E}_{\alpha\beta} = \mathbf{R}\alpha + \mathbf{R}\beta$ (see Exercise 2 above). The Weyl group $\mathscr{W}_{\alpha\beta}$ of $\Phi_{\alpha\beta}$ is generated by the restrictions τ_α, τ_β to $\mathsf{E}_{\alpha\beta}$ of σ_α, σ_β, and $\mathscr{W}_{\alpha\beta}$ may be viewed as a subgroup of $\mathscr{W}$. Prove that the "length" of an element of $\mathscr{W}_{\alpha\beta}$ (relative to τ_α, τ_β) coincides with the length of the corresponding element of $\mathscr{W}$.

9. Prove that there is a unique element σ in $\mathscr{W}$ sending Φ^+ to Φ^- (relative to Δ). Prove that any reduced expression for σ must involve all σ_α ($\alpha \in \Delta$). Discuss $\ell(\sigma)$.

10. Given $\Delta = \{\alpha_1, \ldots, \alpha_l\}$ in Φ, let $\lambda = \sum_{i=1}^{\ell} k_i \alpha_i$ ($k_i \in \mathbf{Z}$, all $k_i \geq 0$ or all $k_i \leq 0$). Prove that either λ is a multiple (possibly 0) of a root, or else there exists $\sigma \in \mathscr{W}$ such that $\sigma\lambda = \sum_{i=1}^{\ell} k_i' \alpha_i$, with some $k_i' > 0$ and some $k_i' < 0$. [Sketch of proof: If λ is not a multiple of any root, then the hyperplane P_λ orthogonal to λ is not included in $\bigcup_{\alpha \in \Phi} P_\alpha$. Take $\mu \in P_\lambda - \bigcup_{\alpha \in \Phi} P_\alpha$. Then find $\sigma \in \mathscr{W}$ for which all $(\alpha_i, \sigma\mu) > 0$. It follows that $0 = (\lambda, \mu) = (\sigma\lambda, \sigma\mu) = \Sigma k(\alpha_i, \sigma\mu)$.]

11. Let Φ be irreducible. Prove that $\Phi^\vee$ is also irreducible. If Φ has all roots of equal length, so does $\Phi^\vee$ (and then $\Phi^\vee$ is isomorphic to Φ). On the other hand, if Φ has two root lengths, then so does $\Phi^\vee$; but if α is long, then $\alpha^\vee$ is short (and vice versa). Use this fact to prove that Φ has a unique maximal short root (relative to the partial order $\prec$ defined by Δ).

12. Let Φ be irreducible. Use Lemmas 10.3B, 10.4C to give an alternate proof that Φ has unique maximal long and short roots,

13. The only reflections in $\mathscr{W}$ are those of the form σ_α $(\alpha \in \Phi)$. [A vector in the reflecting hyperplane would, if orthogonal to no root, be fixed only by the identity in $\mathscr{W}$.]

Notes

The exposition here is an expanded version of that in Serre [2].

11. Classification

In this section Φ denotes a root system of rank ℓ, $\mathscr{W}$ its Weyl group, Δ a base of Φ.

11.1. Cartan matrix of Φ

Fix an ordering $(\alpha_1, \ldots, \alpha_\ell)$ of the simple roots. The matrix $(\langle \alpha_i, \alpha_j \rangle)$ is then called the **Cartan matrix** of Φ. Its entries are called **Cartan integers**. *Examples*: For the systems of rank 2, the matrices are:

$$A_1 \times A_1 \begin{pmatrix} 2 & 0 \\ 0 & 2 \end{pmatrix} ; \ A_2 \begin{pmatrix} 2 & -1 \\ -1 & 2 \end{pmatrix} ; \ B_2 \begin{pmatrix} 2 & -2 \\ -1 & 2 \end{pmatrix} ; \ G_2 \begin{pmatrix} 2 & -1 \\ -3 & 2 \end{pmatrix} .$$

The matrix of course depends on the chosen ordering, but this is not very serious. The important point is that the Cartan matrix is independent of the choice of Δ, thanks to the fact (Theorem 10.3(b)) that $\mathscr{W}$ acts transitively on the collection of bases. The Cartan matrix is nonsingular, as in (8.5), since Δ is a basis of E. It turns out to characterize Φ completely.

Proposition. *Let $\Phi' \subset E'$ be another root system, with base $\Delta' = \{\alpha'_1, \ldots, \alpha'_\ell\}$. If $\langle \alpha_i, \alpha_j \rangle = \langle \alpha'_i, \alpha'_j \rangle$ for $1 \leq i, j \leq \ell$, then the bijection $\alpha_i \mapsto \alpha'_i$ extends (uniquely) to an isomorphism $\phi: E \to E'$ mapping Φ onto Φ' and satisfying $\langle \phi(\alpha), \phi(\beta) \rangle = \langle \alpha, \beta \rangle$ for all $\alpha, \beta \in \Phi$. Therefore, the Cartan matrix of Φ determines Φ up to isomorphism.*

Proof. Since Δ (resp. Δ') is a basis of E (resp. E'), there is a unique vector space isomorphism $\phi: E \to E'$ sending α_i to α'_i $(1 \leq i \leq \ell)$. If $\alpha, \beta \in \Delta$, the hypothesis insures that $\sigma_{\phi(\alpha)}(\phi(\beta)) = \sigma_{\alpha'}(\beta') = \beta' - \langle \beta', \alpha' \rangle \alpha' = \phi(\beta) -$

$\langle \beta, \ \alpha \rangle \ \phi(\alpha) = \phi(\beta - \langle \beta, \ \alpha \rangle \ \alpha) = \phi(\sigma_\alpha(\beta))$. In other words, the following diagram commutes for each $\alpha \in \Delta$:

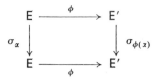

The respective Weyl groups $\mathscr{W}$, $\mathscr{W}'$ are generated by simple reflections (Theorem 10.3(d)), so it follows that the map $\sigma \mapsto \phi \circ \sigma \circ \phi^{-1}$ is an isomorphism of $\mathscr{W}$ onto $\mathscr{W}'$, sending σ_α to $\sigma_{\phi(\alpha)}$ ($\alpha \in \Delta$). But each $\beta \in \Phi$ is conjugate under $\mathscr{W}$ to a simple root (Theorem 10.3(c)), say $\beta = \sigma(\alpha)$ ($\alpha \in \Delta$). This in turn forces $\phi(\beta) = (\phi \circ \sigma \circ \phi^{-1}) \ (\phi(\alpha)) \in \Phi'$. It follows that ϕ maps Φ onto Φ'; moreover, the formula for a reflection shows that ϕ preserves all Cartan integers. ☐

The proposition shows that it is possible in principle to recover Φ from a knowledge of the Cartan integers. In fact, it is not too hard to devise a practical algorithm for writing down all roots (or just all positive roots). Probably the best approach is to consider *root strings* (9.4). Start with the roots of height one, i.e., the simple roots. For any pair $\alpha_i \neq \alpha_j$, the integer r for the α_j-string through α_i is 0 (i.e., $\alpha_i - \alpha_j$ is not a root, thanks to Lemma 10.1), so the integer q equals $-\langle \alpha_i, \ \alpha_j \rangle$. This enables us in particular to write down all roots α of height 2, hence all integers $\langle \alpha, \ \alpha_j \rangle$. For each root α of height 2, the integer r for the α_j-string through α can be determined easily, since α_j can be subtracted at most once (why?), and then q is found, because we know $r - q = \langle \alpha, \ \alpha_j \rangle$. The corollary of Lemma 10.2A assures us that all positive roots are eventually obtained if we repeat this process enough times.

11.2. Coxeter graphs and Dynkin diagrams

If α, β are distinct positive roots, then we know that $\langle \alpha, \ \beta \rangle \langle \beta, \ \alpha \rangle = 0$, 1, 2, or 3 (9.4). Define the **Coxeter graph** of Φ to be a graph having ℓ vertices, the ith joined to the jth ($i \neq j$) by $\langle \alpha_i, \ \alpha_j \rangle \langle \alpha_j, \ \alpha_i \rangle$ edges. *Examples*:

$$A_1 \times A_1 \qquad \circ \qquad \qquad \circ$$
$$A_2 \qquad \circ\!\!-\!\!-\!\!-\!\!-\!\!\circ$$
$$B_2 \qquad \circ\!\!=\!\!=\!\!=\!\!\circ$$
$$G_2 \qquad \circ\!\!\equiv\!\!\equiv\!\!\circ$$

The Coxeter graph determines the numbers $\langle \alpha_i, \ \alpha_j \rangle$ in case all roots have equal length, since then $\langle \alpha_i, \ \alpha_j \rangle = \langle \alpha_j, \ \alpha_i \rangle$. In case more than one root length occurs (e.g., B_2 or G_2), the graph fails to tell us which of a pair of vertices should correspond to a short simple root, which to a long (in case these vertices are joined by two or three edges). (It can, however, be proved that the Coxeter graph determines the Weyl group completely, essentially because it determines the orders of products of generators of $\mathcal{W}$, cf. Exercise 9.3.)

Whenever a double or triple edge occurs in the Coxeter graph of Φ, we can add an arrow pointing to the shorter of the two roots. This additional information allows us to recover the Cartan integers; we call the resulting figure the **Dynkin diagram** of Φ. (As before, this depends on the numbering of simple roots.) For example:

$$B_2 \quad \circ\!\!=\!\!\!\Longrightarrow\!\!\!=\!\!\circ$$
$$G_2 \quad \circ\!\!\equiv\!\!\!\Longleftarrow\!\!\!\equiv\!\!\circ$$

Another example: Given the diagram $\circ\!\!-\!\!-\!\!-\!\!\circ\!\!=\!\!\!\Longrightarrow\!\!\!=\!\!\circ\!\!-\!\!-\!\!-\!\!\circ$ (which turns out to be associated with the root system F_4), the reader can easily recover the Cartan matrix

$$\begin{pmatrix} 2 & -1 & 0 & 0 \\ -1 & 2 & -2 & 0 \\ 0 & -1 & 2 & -1 \\ 0 & 0 & -1 & 2 \end{pmatrix}.$$

11.3. Irreducible components

Recall (10.4) that Φ is irreducible if and only if Φ (or, equivalently, Δ) cannot be partitioned into two proper, orthogonal subsets. It is clear that Φ is *irreducible if and only if its Coxeter graph is connected* (in the usual sense). In general, there will be a number of connected components of the Coxeter graph; let $\Delta = \Delta_1 \cup \ldots \cup \Delta_t$ be the corresponding partition of Δ into mutually orthogonal subsets. If E_i is the span of Δ_i, it is clear that $E = E_1 \oplus \ldots \oplus E_t$ (orthogonal direct sum). Moreover, the Z-linear combinations of Δ_i which are roots (call this set Φ_i) obviously form a root system in E_i, whose Weyl group is the restriction to E_i of the subgroup of $\mathcal{W}$ generated by all σ_α ($\alpha \in \Delta_i$). Finally, each E_i is $\mathcal{W}$-invariant (since $\alpha \notin \Delta_i$ implies that σ_α acts trivially on E_i), so the (easy) argument required for Exercise 9.1 shows immediately that each root lies in one of the E_i, i.e., $\Phi = \Phi_1 \cup \ldots \cup \Phi_t$.

Proposition. Φ *decomposes (uniquely) as the union of irreducible root systems* Ψ_i *(in subspaces E_i of E) such that* $E = E_1 \oplus \ldots \oplus E_t$ *(orthogonal direct sum).* □

11.4. Classification theorem

The discussion in (11.3) shows that it is sufficient to classify the irreducible root systems, or equivalently, the connected Dynkin diagrams (cf. Proposition 11.1).

Theorem. *If Φ is an irreducible root system of rank ℓ, its Dynkin diagram is one of the following (ℓ vertices in each case):*

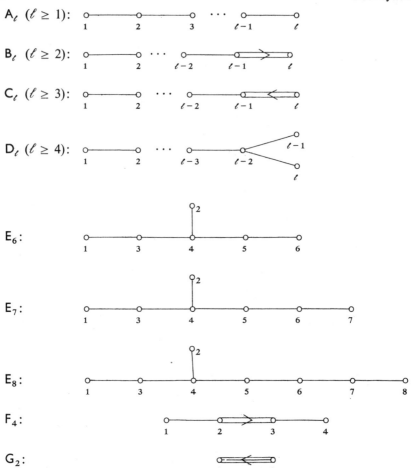

The restrictions on ℓ for types $A_\ell - D_\ell$ are imposed in order to avoid duplication. Relative to the indicated numbering of simple roots, the corresponding Cartan matrices are given in Table 1. Inspection of the diagrams listed above reveals that in all cases except B_ℓ, C_ℓ, the Dynkin diagram can be deduced from the Coxeter graph. However, B_ℓ and C_ℓ both come from a single Coxeter graph, and differ in the relative numbers of short and long simple roots. (These root systems are actually dual to each other, cf. Exercise 5.)

   Proof of Theorem. The idea of the proof is to classify first the possible Coxeter graphs (ignoring relative lengths of roots), then see what Dynkin diagrams result. Therefore, we shall merely apply some elementary euclidean geometry to finite sets of vectors whose pairwise angles are those prescribed by the Coxeter graph. Since we are ignoring lengths, it is easier to work for the time being with sets of unit vectors. For maximum flexibility, we make

Table 1. *Cartan matrices*

$$A_\ell: \begin{pmatrix} 2 & -1 & 0 & & \cdot & \cdot & \cdot & & 0 \\ -1 & 2 & -1 & 0 & \cdot & \cdot & \cdot & & 0 \\ 0 & -1 & 2 & -1 & 0 & \cdot & \cdot & \cdot & 0 \\ \cdot & & \cdot & \cdot & \cdot & & & & \cdot \\ 0 & 0 & 0 & 0 & & \cdot & \cdot & -1 & 2 \end{pmatrix}$$

$$B_\ell: \begin{pmatrix} 2 & -1 & 0 & & \cdot & \cdot & \cdot & & 0 \\ -1 & 2 & -1 & 0 & \cdot & \cdot & \cdot & & 0 \\ \cdot & & \cdot & \cdot & \cdot & & \cdot & & \cdot \\ 0 & 0 & 0 & & \cdot & \cdot & \cdot & -1 & 2 & -2 \\ 0 & 0 & 0 & & \cdot & \cdot & \cdot & 0 & -1 & 2 \end{pmatrix}$$

$$C_\ell: \begin{pmatrix} 2 & -1 & 0 & & \cdot & \cdot & \cdot & & 0 \\ -1 & 2 & -1 & & \cdot & \cdot & \cdot & & 0 \\ 0 & -1 & 2 & -1 & \cdot & \cdot & \cdot & & 0 \\ \cdot & & \cdot & \cdot & & & \cdot & & \cdot \\ 0 & 0 & 0 & & \cdot & \cdot & \cdot & -1 & 2 & -1 \\ 0 & 0 & & & \cdot & \cdot & \cdot & 0 & -2 & 2 \end{pmatrix}$$

$$D_\ell: \begin{pmatrix} 2 & -1 & 0 & & \cdot & \cdot & \cdot & & & 0 \\ -1 & 2 & -1 & & \cdot & \cdot & \cdot & & & 0 \\ \cdot & \cdot & & \cdot & & \cdot & & & & \cdot \\ 0 & 0 & & \cdot & \cdot & -1 & 2 & -1 & 0 & 0 \\ 0 & 0 & & \cdot & \cdot & & -1 & 2 & -1 & -1 \\ 0 & 0 & & \cdot & \cdot & & 0 & -1 & 2 & 0 \\ 0 & 0 & & \cdot & \cdot & & 0 & -1 & 0 & 2 \end{pmatrix}$$

$$E_6: \begin{pmatrix} 2 & 0 & -1 & 0 & 0 & 0 \\ 0 & 2 & 0 & -1 & 0 & 0 \\ -1 & 0 & 2 & -1 & 0 & 0 \\ 0 & -1 & -1 & 2 & -1 & 0 \\ 0 & 0 & 0 & -1 & 2 & -1 \\ 0 & 0 & 0 & 0 & -1 & 2 \end{pmatrix}$$

$$E_7: \begin{pmatrix} 2 & 0 & -1 & 0 & 0 & 0 & 0 \\ 0 & 2 & 0 & -1 & 0 & 0 & 0 \\ -1 & 0 & 2 & -1 & 0 & 0 & 0 \\ 0 & -1 & -1 & 2 & -1 & 0 & 0 \\ 0 & 0 & 0 & -1 & 2 & -1 & 0 \\ 0 & 0 & 0 & 0 & -1 & 2 & -1 \\ 0 & 0 & 0 & 0 & 0 & -1 & 2 \end{pmatrix}$$

$$E_8: \begin{pmatrix} 2 & 0 & -1 & 0 & 0 & 0 & 0 & 0 \\ 0 & 2 & 0 & -1 & 0 & 0 & 0 & 0 \\ -1 & 0 & 2 & -1 & 0 & 0 & 0 & 0 \\ 0 & -1 & -1 & 2 & -1 & 0 & 0 & 0 \\ 0 & 0 & 0 & -1 & 2 & -1 & 0 & 0 \\ 0 & 0 & 0 & 0 & -1 & 2 & -1 & 0 \\ 0 & 0 & 0 & 0 & 0 & -1 & 2 & -1 \\ 0 & 0 & 0 & 0 & 0 & 0 & -1 & 2 \end{pmatrix}$$

$$F_4: \begin{pmatrix} 2 & -1 & 0 & 0 \\ -1 & 2 & -2 & 0 \\ 0 & -1 & 2 & -1 \\ 0 & 0 & -1 & 2 \end{pmatrix} \qquad G_2: \begin{pmatrix} 2 & -1 \\ -3 & 2 \end{pmatrix}$$

only the following assumptions: E is a euclidean space (of arbitrary dimension), $\mathfrak{A} = \{\varepsilon_1, \ldots, \varepsilon_n\}$ is a set of n linearly independent unit vectors which satisfy $(\varepsilon_i, \varepsilon_j) \leq 0$ $(i \neq j)$ and $4(\varepsilon_i, \varepsilon_j)^2 = 0, 1, 2,$ or 3 $(i \neq j)$. Such a set of vectors is called (for brevity) **admissible**. (*Example*: Elements of a base for a root system, each divided by its length.) We attach a graph Γ to the set $\mathfrak{A}$ just as we did above to the simple roots in a root system, with vertices i and j $(i \neq j)$ joined by $4(\varepsilon_i, \varepsilon_j)^2$ edges. Now our task is to determine all the connected graphs associated with admissible sets of vectors (these include all connected Coxeter graphs). This we do in steps, the first of which is obvious. (Γ is not assumed to be connected until later on.)

(1) *If some of the ε_i are discarded, the remaining ones still form an admissible set, whose graph is obtained from Γ by omitting the corresponding vertices and all incident edges.*

(2) *The number of pairs of vertices in Γ connected by at least one edge is strictly less than n.* Set $\varepsilon = \sum_{i=1}^{n} \varepsilon_i$. Since the ε_i are linearly independent, $\varepsilon \neq 0$. So $0 < (\varepsilon, \varepsilon) = n + 2 \sum_{i<j} (\varepsilon_i, \varepsilon_j)$. Let i, j be a pair of (distinct) indices for which $(\varepsilon_i, \varepsilon_j) \neq 0$ (i.e., let vertices i and j be joined). Then $4(\varepsilon_i, \varepsilon_j)^2 = 1, 2,$ or 3, so in particular $2(\varepsilon_i, \varepsilon_j) \leq -1$. In view of the above inequality, the number of such pairs cannot exceed $n-1$.

(3) Γ *contains no cycles.* A cycle would be the graph Γ' of an admissible subset $\mathfrak{A}'$ of $\mathfrak{A}$ (cf. (1)), and then Γ' would violate (2), with n replaced by Card $\mathfrak{A}'$.

(4) *No more than three edges can originate at a given vertex of Γ.* Say $\varepsilon \in \mathfrak{A}$, and $\eta_1, \ldots, \eta_k$ are the vectors in $\mathfrak{A}$ connected to ε (by 1, 2, or 3 edges each), i.e., $(\varepsilon, \eta_i) < 0$ with $\varepsilon, \eta_1, \ldots, \eta_k$ all distinct. In view of (3), no two η's can be connected, so $(\eta_i, \eta_j) = 0$ for $i \neq j$. Because $\mathfrak{A}$ is linearly independent, some unit vector η_0 in the span of $\varepsilon, \eta_1, \ldots, \eta_k$ is orthogonal to $\eta_1, \ldots, \eta_k$; clearly $(\varepsilon, \eta_0) \neq 0$ for such η_0. Now $\varepsilon = \sum_{i=0}^{k} (\varepsilon, \eta_i)\eta_i$, so $1 = (\varepsilon, \varepsilon) = \sum_{i=0}^{k} (\varepsilon, \eta_i)^2$. This forces $\sum_{i=1}^{k} (\varepsilon, \eta_i)^2 < 1$, or $\sum_{i=1}^{k} 4(\varepsilon, \eta_i)^2 < 4$. But $4(\varepsilon, \eta_i)^2$ is the number of edges joining ε to η_i in Γ.

(5) *The only connected graph Γ of an admissible set $\mathfrak{A}$ which can contain a triple edge is* ⊶⊶⊶ (*the Coxeter graph* G_2). This follows at once from (4).

(6) *Let $\{\varepsilon_1, \ldots, \varepsilon_k\} \subset \mathfrak{A}$ have subgraph* ○———○·····○———○ (*a simple chain in Γ. If $\mathfrak{A}' = (\mathfrak{A} - \{\varepsilon_1, \ldots, \varepsilon_k\}) \cup \{\varepsilon\}$, $\varepsilon = \sum_{i=1}^{k} \varepsilon_i$, then $\mathfrak{A}'$ is admissible.*

(The graph of $\mathfrak{A}'$ is obtained from Γ by shrinking the simple chain to a point.) Linear independence of $\mathfrak{A}'$ is obvious. By hypothesis, $2(\varepsilon_i, \varepsilon_{i+1}) = -1$ $(1 \leq i \leq k-1)$, so $(\varepsilon, \varepsilon) = k + 2 \sum_{i<j} (\varepsilon_i, \varepsilon_j) = k - (k-1) = 1$. So ε is a unit vector. Any $\eta \in \mathfrak{A} - \{\varepsilon_1, \ldots, \varepsilon_k\}$ can be connected to at most one of $\varepsilon_1, \ldots, \varepsilon_k$ (by (3)), so $(\eta, \varepsilon) = 0$ or else $(\eta, \varepsilon) = (\eta, \varepsilon_i)$ for $1 \leq i \leq k$. In either case, $4(\eta, \varepsilon)^2 = 0, 1, 2,$ or 3.

(7) Γ *contains no subgraph of the form*:

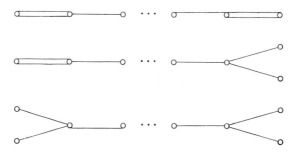

Suppose one of these graphs occurred in Γ; by (1) it would be the graph of an admissible set. But (6) allows us to replace the simple chain in each case by a single vertex, yielding (respectively) the following graphs which violate (4):

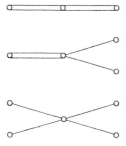

(8) *Any connected graph Γ of an admissible set has one of the following forms*:

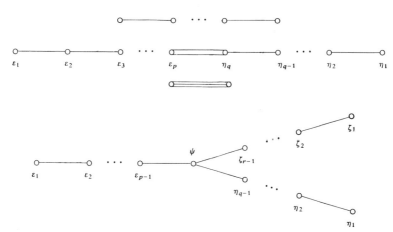

Indeed, only  contains a triple edge, by (5). A connected graph containing more than one double edge would contain a subgraph which (7) forbids, so at most one double edge occurs. Moreover, if Γ has a double edge, it cannot also have a "node" (branch point)

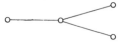

(again by (7)), so the second graph pictured is the only possibility (cycles being forbidden by (3)). Finally, let Γ have only single edges; if Γ has no node, it must be a simple chain (again because no cycles are allowed). It cannot contain more than one node (7), so the fourth graph is the only remaining possibility.

(9) *The only connected Γ of the second type in* (8) *is the Coxeter graph* F_4

o————o=======o————o *or the Coxeter graph* $\mathsf{B}_n(=\mathsf{C}_n)$ o————o $\cdots$

o————o=======o .

Set $\varepsilon = \sum_{i=1}^{p} i\varepsilon_i$, $\eta = \sum_{i=1}^{q} i\eta_i$. By hypothesis, $2(\varepsilon_i, \varepsilon_{i+1}) = -1 = 2(\eta_i, \eta_{i+1})$, and other pairs are orthogonal, so $(\varepsilon, \varepsilon) = \sum_{i=1}^{p} i^2 - \sum_{i=1}^{p-1} i(i+1) = p(p+1)/2$, (η, η) $= q(q+1)/2$. Since $4(\varepsilon_p, \eta_q)^2 = 2$, we also have $(\varepsilon, \eta)^2 = p^2 q^2 (\varepsilon_p, \eta_q)^2 = p^2 q^2 / 2$. The Schwartz inequality implies (since ε, η are obviously independent) that $(\varepsilon, \eta)^2 < (\varepsilon, \varepsilon)(\eta, \eta)$, or $p^2 q^2 / 2 < p(p+1)q(q+1)/4$, whence $(p-1)(q-1)$ < 2. The possibilities are: $p = q = 2$ (whence F_4) or $p = 1$ (q arbitrary), $q = 1$ (p arbitrary).

(10) *The only connected Γ of the fourth type in* (8) *is the Coxeter graph* D_n

o————o $\cdots$ o————o $\prec$ or the *Coxeter graph* E_n ($n = 6, 7$ or 8)

o————o————o————o $\cdots$ o. Set $\varepsilon = \Sigma\, i\varepsilon_i$, $\eta = \Sigma\, i\eta_i$, $\zeta = \Sigma\, i\zeta_i$. It is clear that ε, η, ζ are mutually orthogonal, linearly independent vectors, and that ψ is not in their span. As in the proof of (4) we therefore obtain $\cos^2 \theta_1 + \cos^2 \theta_2 + \cos^2 \theta_3 < 1$, where $\theta_1, \theta_2, \theta_3$ are the respective angles between ψ and ε, η, ζ. The same calculation as in (9), with $p-1$ in place of p, shows that $(\varepsilon, \varepsilon) = p(p-1)/2$, and similarly for η, ζ. Therefore $\cos^2 \theta_1 = (\varepsilon, \psi)^2/(\varepsilon, \varepsilon)(\psi, \psi) = (p-1)^2(\varepsilon_{p-1}, \psi)^2/(\varepsilon, \varepsilon) = \frac{1}{4}(2(p-1)^2/p(p-1)) = (p-1)/2p = \frac{1}{2}(1 - 1/p)$. Similarly for θ_2, θ_3. Adding, we get the inequality $\frac{1}{2}(1 - 1/p + 1 - 1/q + 1 - 1/r) < 1$, or (*) $1/p + 1/q + 1/r > 1$. (This inequality, by the way, has a long mathematical history.) By changing labels we may assume that $1/p \le 1/q \le 1/r$ ($\le 1/2$; if p, q, or r equals 1, we are back in type A_n). In particular, the inequality (*) implies $3/2 \ge 3/r > 1$, so $r = 2$. Then $1/p + 1/q > 1/2$, $2/q > 1/2$, and $2 \le q < 4$. If $q = 3$, then $1/p > 1/6$ and necessarily $p < 6$. So the possible triples (p, q, r) turn out to be: $(p, 2, 2)$ $= \mathsf{D}_n$; $(3, 3, 2) = \mathsf{E}_6$; $(4, 3, 2) = \mathsf{E}_7$; $(5, 3, 2) = \mathsf{E}_8$.

The preceding argument shows that the connected graphs of admissible sets of vectors in euclidean space are all to be found among the Coxeter graphs of types A–G. In particular, the Coxeter graph of a root system must be of one of these types. But in all cases except B_ℓ, C_ℓ, the Coxeter graph uniquely determines the Dynkin diagram, as remarked at the outset. So the theorem follows. □

Exercises

1. Verify the Cartan matrices (Table 1).
2. Calculate the determinants of the Cartan matrices (using induction on ℓ for types A_ℓ–D_ℓ), which are as follows:

   A_ℓ: $\ell + 1$; B_ℓ: 2; C_ℓ: 2; D_ℓ: 4; E_6: 3; E_7: 2; E_8, F_4 and G_2: 1.

3. Use the algorithm of (11.1) to write down all roots for G_2. Do the same for C_3: $\begin{pmatrix} 2 & -1 & 0 \\ -1 & 2 & -1 \\ 0 & -2 & 2 \end{pmatrix}$.

4. Prove that the Weyl group of a root system Φ is isomorphic to the direct product of the respective Weyl groups of its irreducible components.
5. Prove that each irreducible root system is isomorphic to its dual, except that B_ℓ, C_ℓ are dual to each other.
6. Prove that an inclusion of one Dynkin diagram in another (e.g., E_6 in E_7 or E_7 in E_8) induces an inclusion of the corresponding root systems.

Notes

Our proof of the classification theorem follows Jacobson [1]. For a somewhat different approach, see Carter [1]. Bourbaki [2] emphasizes the classification of Coxeter groups, of which the Weyl groups of root systems are important examples.

12. Construction of root systems and automorphisms

In §11 the possible (connected) Dynkin diagrams of (irreducible) root systems were all determined. It remains to be shown that each diagram of type A–G does in fact belong to a root system Φ. Afterwards we shall briefly discuss Aut Φ. The existence of root systems of type $A_\ell - D_\ell$ could actually be shown by verifying for each classical linear Lie algebra (1.2) that its root system is of the indicated type, which of course, requires that we first prove the semisimplicity of these algebras (cf. §19). But it is easy enough to give a direct construction of the root system, which moreover makes plain the structure of its Weyl group.

12.1. Construction of types A–G

We shall work in various spaces $\mathbf{R}^n$, where the inner product is the usual one and where $\varepsilon_1, \ldots, \varepsilon_n$ denote the usual orthonormal unit vectors which form a basis of $\mathbf{R}^n$. The $\mathbf{Z}$-span of this basis is (by definition) a **lattice**, denoted I. In each case we shall take E to be $\mathbf{R}^n$ (or a suitable subspace thereof, with the inherited inner product). Then Φ will be defined to be the set of all vectors in I (or a suitable subgroup thereof) having specified length or lengths.

Since I is a lattice (hence discrete in the usual topology of $\mathbf{R}^n$), while the set of vectors in $\mathbf{R}^n$ having one or two given lengths is compact (closed and bounded), Φ is then obviously finite, and will exclude 0 by definition. In each case it will be evident that Φ spans E (indeed, a base of Φ will be exhibited explicitly). Therefore (R1) is satisfied. The choice of lengths will also make it obvious that (R2) holds. For (R3) it is enough to check that the reflection σ_α ($\alpha \in \Phi$) maps Φ back into I (or its specified subgroup), since then $\sigma_\alpha(\Phi)$ automatically consists of vectors of the required lengths. But then (R3) follows from (R4). As to (R4), it suffices to choose squared lengths dividing 2, since it is automatic that all inner products $(\alpha, \beta) \in \mathbf{Z}$ ($\alpha, \beta \in I$).

Having made these preliminary remarks, we now treat the separate cases $\mathsf{A} - \mathsf{G}$. After verifying (R1) to (R4) in the way just sketched, the reader should observe that the resulting Cartan matrix matches that in Table 1 (11.4).

A_ℓ ($\ell \ge 1$): Let E be the ℓ-dimensional subspace of $\mathbf{R}^{\ell+1}$ orthogonal to the vector $\varepsilon_1 + \ldots + \varepsilon_{\ell+1}$. Let $I' = I \cap \mathsf{E}$, and take Φ to be the set of all vectors $\alpha \in I'$ for which $(\alpha, \alpha) = 2$. It is obvious that $\Phi = \{\varepsilon_i - \varepsilon_j, i \ne j\}$. The vectors $\alpha_i = \varepsilon_i - \varepsilon_{i+1}$ ($1 \le i \le \ell$) are independent, and $\varepsilon_i - \varepsilon_j = (\varepsilon_i - \varepsilon_{i+1}) + (\varepsilon_{i+1} - \varepsilon_{i+2}) + \ldots + (\varepsilon_{j-1} - \varepsilon_j)$ if $i < j$, which shows that they form a base of Φ. It is clear that the Cartan matrix A_ℓ results. Finally, notice that the reflection with respect to α_i permutes the subscripts $i, i+1$ and leaves all other subscripts fixed. Thus σ_{α_i} corresponds to the transposition $(i, i+1)$ in the symmetric group $\mathscr{S}_{\ell+1}$; these transpositions generate $\mathscr{S}_{\ell+1}$, so we obtain a natural isomorphism of $\mathscr{W}$ onto $\mathscr{S}_{\ell+1}$.

B_ℓ ($\ell \ge 2$): Let $\mathsf{E} = \mathbf{R}^\ell$, $\Phi = \{\alpha \in I | (\alpha, \alpha) = 1 \text{ or } 2\}$. It is easy to check that Φ consists of the vectors $\pm \varepsilon_i$ (of squared length 1) and the vectors $\pm (\varepsilon_i \pm \varepsilon_j)$, $i \ne j$ (of squared length 2). The ℓ vectors $\varepsilon_1 - \varepsilon_2, \varepsilon_2 - \varepsilon_3, \ldots, \varepsilon_{\ell-1} - \varepsilon_\ell, \varepsilon_\ell$ are independent; a short root $\varepsilon_i = (\varepsilon_i - \varepsilon_{i+1}) + (\varepsilon_{i+1} - \varepsilon_{i+2}) + \ldots + (\varepsilon_{\ell-1} - \varepsilon_\ell) + \varepsilon_\ell$, while a long root $\varepsilon_i - \varepsilon_j$ or $\varepsilon_i + \varepsilon_j$ is similarly expressible. The Cartan matrix for this (ordered) base is clearly B_ℓ. $\mathscr{W}$ acts as the group of all permutations and sign changes of the set $\{\varepsilon_1, \ldots, \varepsilon_\ell\}$, so $\mathscr{W}$ is isomorphic to the semidirect product of $(\mathbf{Z}/2\mathbf{Z})^\ell$ and $\mathscr{S}_\ell$ (the latter acting on the former).

C_ℓ ($\ell \ge 3$): C_ℓ ($\ell \ge 2$) may be viewed most conveniently as the root system dual to B_ℓ (with $\mathsf{B}_2 = \mathsf{C}_2$), cf. Exercise 11.5. The reader can verify directly that in $\mathsf{E} = \mathbf{R}^\ell$, the set of all $\pm 2\varepsilon_i$ and all $\pm (\varepsilon_i \pm \varepsilon_j)$, $i \ne j$, forms a root system of type C_ℓ, with base $(\varepsilon_1 - \varepsilon_2, \ldots, \varepsilon_{\ell-1} - \varepsilon_\ell, 2\varepsilon_\ell)$. Of course the Weyl group is isomorphic to that of B_ℓ.

D_ℓ ($\ell \ge 4$): Let $\mathsf{E} = \mathbf{R}^\ell$, $\Phi = \{\alpha \in I | (\alpha, \alpha) = 2\} = \{\pm (\varepsilon_i \pm \varepsilon_j), i \ne j\}$. For a base take the ℓ independent vectors $\varepsilon_1 - \varepsilon_2, \ldots, \varepsilon_{\ell-1} - \varepsilon_\ell, \varepsilon_{\ell-1} + \varepsilon_\ell$ (so D_ℓ results). The Weyl group is the group of permutations and sign changes involving only *even numbers of signs* of the set $\{\varepsilon_1, \ldots, \varepsilon_\ell\}$. So $\mathscr{W}$ is isomorphic to the semidirect product of $(\mathbf{Z}/2\mathbf{Z})^{\ell-1}$ and $\mathscr{S}_\ell$.

E_6, E_7, E_8: We know that E_6, E_7 can be identified canonically with subsystems of E_8 (Exercise 11.6), so it suffices to construct E_8. This is slightly complicated. Take $E = \mathbf{R}^8$, $I' = I + \mathbf{Z}((\varepsilon_1 + \ldots + \varepsilon_8)/2)$, $I'' =$ subgroup of I' consisting of all elements $\Sigma c_i \varepsilon_i + \dfrac{c}{2}(\varepsilon_1 + \ldots + \varepsilon_8)$ for which $c + \Sigma c_i$ is an *even* integer. (Check that this is a subgroup!) Define $\Phi = \{\alpha \in I'' | (\alpha, \alpha) = 2\}$. It is easy to see that Φ consists of the obvious vectors $\pm(\varepsilon_i \pm \varepsilon_j)$, $i \neq j$, along with the less obvious ones $\frac{1}{2} \sum_{i=1}^{8} (-1)^{k(i)} \varepsilon_i$ (where the $k(i) = 0$, 1, add up to an even integer). By inspection, all inner products here are in $\mathbf{Z}$ (this has to be checked, because we are working in a larger lattice than I). As a base we take $\{\frac{1}{2}(\varepsilon_1 + \varepsilon_8 - (\varepsilon_2 + \ldots + \varepsilon_7)), \; \varepsilon_1 + \varepsilon_2, \; \varepsilon_2 - \varepsilon_1, \; \varepsilon_3 - \varepsilon_2, \; \varepsilon_4 - \varepsilon_3, \; \varepsilon_5 - \varepsilon_4, \; \varepsilon_6 - \varepsilon_5, \; \varepsilon_7 - \varepsilon_6\}$. (This has been ordered so as to correspond to the Cartan matrix for E_8 in Table 1 (11.4).) The reader is invited to contemplate for himself the action of the Weyl group, whose order can be shown to be $2^{14} 3^5 5^2 7$.

F_4: Let $E = \mathbf{R}^4$, $I' = I + \mathbf{Z}((\varepsilon_1 + \varepsilon_2 + \varepsilon_3 + \varepsilon_4)/2)$, $\Phi = \{\alpha \in I' | (\alpha, \alpha) = 1 \text{ or } 2\}$. Then Φ consists of all $\pm \varepsilon_i$, all $\pm(\varepsilon_i - \varepsilon_j)$, $i \neq j$, as well as all $\pm \frac{1}{2}(\varepsilon_1 \pm \varepsilon_2 \pm \varepsilon_3 \pm \varepsilon_4)$, where the signs may be chosen independently. By inspection, all inner products are integral. As a base take $\{\varepsilon_2 - \varepsilon_3, \; \varepsilon_3 - \varepsilon_4, \; \varepsilon_4, \; \frac{1}{2}(\varepsilon_1 - \varepsilon_2 - \varepsilon_3 - \varepsilon_4)\}$. Here $\mathscr{W}$ has order 1152.

G_2: We already constructed G_2 explicitly in §9. Abstractly, we can take E to be the subspace of $\mathbf{R}^3$ orthogonal to $\varepsilon_1 + \varepsilon_2 + \varepsilon_3$, $I' = I \cap E$, $\Phi = \{\alpha \in I' | (\alpha, \alpha) = 2 \text{ or } 6\}$. So $\Phi = \pm \{\varepsilon_1 - \varepsilon_2, \; \varepsilon_2 - \varepsilon_3, \; \varepsilon_1 - \varepsilon_3, \; 2\varepsilon_1 - \varepsilon_2 - \varepsilon_3, \; 2\varepsilon_2 - \varepsilon_1 - \varepsilon_3, \; 2\varepsilon_3 - \varepsilon_1 - \varepsilon_2\}$. As a base choose $\varepsilon_1 - \varepsilon_2$, $-2\varepsilon_1 + \varepsilon_2 + \varepsilon_3$. (How does $\mathscr{W}$ act?)

Theorem. *For each Dynkin diagram (or Cartan matrix) of type* A–G, *there exists an irreducible root system having the given diagram.* □

12.2. Automorphisms of Φ

We are going to give a complete description of Aut Φ, for each root system Φ. Recall that Lemma 9.2 implies that $\mathscr{W}$ is a normal subgroup of Aut Φ (Exercise 9.6). Let $\Gamma = \{\sigma \in \text{Aut } \Phi | \sigma(\Delta) = \Delta\}$, Δ a fixed base of Φ. Evidently, Γ is a subgroup of Aut Φ. If $\tau \in \Gamma \cap \mathscr{W}$, then $\tau = 1$ by virtue of the simple transitivity of $\mathscr{W}$ (Theorem 10.3(e)). Moreover, if $\tau \in \text{Aut } \Phi$ is arbitrary, then $\tau(\Delta)$ is evidently another base of Φ, so there exists $\sigma \in \mathscr{W}$ such that $\sigma\tau(\Delta) = \Delta$ (Theorem 10.3(b)), whence $\tau \in \Gamma\mathscr{W}$. It follows that *Aut* Φ *is the semidirect product of* Γ *and* $\mathscr{W}$.

For all $\tau \in \text{Aut } \Phi$, all $\alpha, \beta \in \Phi$, we have $\langle \alpha, \beta \rangle = \langle \tau(\alpha), \tau(\beta) \rangle$. Therefore, each $\tau \in \Gamma$ determines an automorphism (in the obvious sense) of the Dynkin diagram of Φ. If τ acts trivially on the diagram, then $\tau = 1$ (because Δ spans E). On the other hand, each automorphism of the Dynkin diagram obviously determines an automorphism of Φ (cf. Proposition 11.1). So Γ *may be*

Dim

Table 1.

Type	Number of Positive Roots	Order of $\mathcal{W}$	Structure of $\mathcal{W}$	Γ
A_ℓ	$\binom{\ell+1}{2}$	$(\ell+1)!$	$\mathscr{S}_{\ell+1}$	$\mathbf{Z}/2\mathbf{Z}\ (\ell \geq 2)$
B_ℓ, C_ℓ	ℓ^2	$2^\ell \ell!$	$(\mathbf{Z}/2\mathbf{Z})^\ell \rtimes \mathscr{S}_\ell$	1
D_ℓ	$\ell^2 - \ell$	$2^{\ell-1}\ell!$	$(\mathbf{Z}/2\mathbf{Z})^{\ell-1} \rtimes \mathscr{S}_\ell$	$\begin{cases}\mathscr{S}_3\ (\ell = 4)\\\mathbf{Z}/2\mathbf{Z}\ (\ell > 4)\end{cases}$
E_6	36	$27\ 3^4\ 5$		$\mathbf{Z}/2\mathbf{Z}$
E_7	63	$2^{10}\ 3^4\ 5\ 7$		1
E_8	120	$2^{14}\ 3^5\ 5^2\ 7$		1
F_4	24	$2^7\ 3^2$		1
G_2	6	$2^2\ 3$	$\mathscr{D}_6$	1

identified with the group of **diagram automorphisms**. A glance at the list in (11.4) yields a description of Γ, summarized in Table 1 along with other useful data, for Φ irreducible. (Since diagram automorphisms other than the identity exist only in cases of single root length, when the Dynkin diagram and Coxeter graph coincide, the term **graph automorphism** may also be used.)

Exercises

1. Verify the details of the constructions in (12.1).
2. Verify Table 2.
3. Let $\Phi \subset E$ satisfy (R1), (R3), (R4), but not (R2), cf. Exercise 9.9. Suppose moreover that Φ is irreducible, in the sense of §11. Prove that Φ is the union of root systems of type B_n, C_n in E ($n = \dim E$), where the long roots of B_n are also the short roots of C_n. (This is called the *non-reduced root system* of type BC_n in the literature.)

Table 2. *Highest long and short roots*

Type	Long	Short
A_ℓ	$\alpha_1 + \alpha_2 + \ldots + \alpha_\ell$	
B_ℓ	$\alpha_1 + 2\alpha_2 + 2\alpha_3 + \ldots + 2\alpha_\ell$	$\alpha_1 + \alpha_2 + \ldots + \alpha_\ell$
C_ℓ	$2\alpha_1 + 2\alpha_2 + \ldots + 2\alpha_{\ell-1} + \alpha_\ell$	$\alpha_1 + 2\alpha_2 + \ldots + 2\alpha_{\ell-1} + \alpha_\ell$
D_ℓ	$\alpha_1 + 2\alpha_2 + \ldots + 2\alpha_{\ell-2} + \alpha_{\ell-1} + \alpha_\ell$	
E_6	$\alpha_1 + 2\alpha_2 + 2\alpha_3 + 3\alpha_4 + 2\alpha_5 + \alpha_6$	
E_7	$2\alpha_1 + 2\alpha_2 + 3\alpha_3 + 4\alpha_4 + 3\alpha_5 + 2\alpha_6 + \alpha_7$	
E_8	$2\alpha_1 + 3\alpha_2 + 4\alpha_3 + 6\alpha_4 + 5\alpha_5 + 4\alpha_6 + 3\alpha_7 + 2\alpha_8$	
F_4	$2\alpha_1 + 3\alpha_2 + 4\alpha_3 + 2\alpha_4$	$\alpha_1 + 2\alpha_2 + 3\alpha_3 + 2\alpha_4$
G_2	$3\alpha_1 + 2\alpha_2$	$2\alpha_1 + \alpha_2$

4. Prove that the long roots in G_2 form a root system in E of type A_2.
5. In constructing C_ℓ, would it be correct to characterize Φ as the set of all vectors in I of squared length 2 or 4? Explain.
6. Prove that the map $\alpha \mapsto -\alpha$ is an automorphism of Φ. Try to decide for which irreducible Φ this belongs to the Weyl group.
7. Describe Aut Φ when Φ is not irreducible.

Notes

The treatment here follows Serre [2]. More information about the individual root systems may be found in Bourbaki [2].

13. Abstract theory of weights

In this section we describe that part of the representation theory of semisimple Lie algebras which depends only on the root system. (None of this is needed until Chapter VI.) Let Φ be a root system in a euclidean space E, with Weyl group $\mathscr{W}$.

13.1. Weights

Let Λ be the set of all $\lambda \in E$ for which $\langle \lambda, \alpha \rangle \in \mathbf{Z}$ $(\alpha \in \Phi)$, and call its elements **weights**. Since $\langle \lambda, \alpha \rangle = \dfrac{2(\lambda, \alpha)}{(\alpha, \alpha)}$ depends linearly on λ, Λ is a subgroup of E including Φ. Denote by Λ_r the **root lattice** (=subgroup of Λ generated by Φ). Λ_r is a lattice in E in the technical sense: it is the **Z**-span of an **R**-basis of E (namely, any set of simple roots). Fix a base $\Delta \subset \Phi$, and define $\lambda \in \Lambda$ to be **dominant** if all the integers $\langle \lambda, \alpha \rangle$ $(\alpha \in \Delta)$ are nonnegative, **strongly dominant** if these integers are positive. Let Λ^+ be the set of all dominant weights. In the language of (10.1), Λ^+ is the set of all weights lying in the closure of the fundamental Weyl chamber $\mathfrak{C}(\Delta)$, while $\Lambda \cap \mathfrak{C}(\Delta)$ is the set of strongly dominant weights.

If $\Delta = \{\alpha_1, \ldots, \alpha_\ell\}$, then the vectors $2\alpha_i/(\alpha_i, \alpha_i)$ again form a basis of E. Let $\lambda_1, \ldots, \lambda_\ell$ be the dual basis (relative to the inner product on E): $\dfrac{2(\lambda_i, \alpha_j)}{(\alpha_j, \alpha_j)} = \delta_{ij}$. Since all $\langle \lambda_i, \alpha \rangle$ $(\alpha \in \Delta)$ are nonnegative integers, the λ_i are dominant weights. We call them the **fundamental dominant weights (relative to Δ)**. Notice that $\sigma_i \lambda_j = \lambda_j - \delta_{ij} \alpha_i$. If $\lambda \in E$ is arbitrary, e.g., any weight, let $m_i = \langle \lambda, \alpha_i \rangle$. Then $0 = \langle \lambda - \Sigma m_i \lambda_i, \alpha \rangle$ for each simple root α, which implies that $(\lambda - \Sigma m_i \lambda_i, \alpha) = 0$ as well, or that $\lambda = \Sigma m_i \lambda_i$. Therefore, Λ *is a lattice with basis* $(\lambda_i, 1 \leq i \leq \ell)$, *and* $\lambda \in \Lambda^+$ *if and only if all* $m_i \geq 0$. (Cf. Figure 1, for type A_2.)

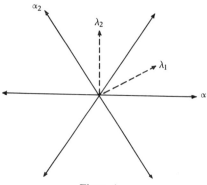

Figure 1

It is an elementary fact about lattices that Λ/Λ_r must be a finite group (called the **fundamental group** of Φ). We can see this directly as follows. Write $\alpha_i = \sum_j m_{ij}\lambda_j \; (m_{ij} \in \mathbf{Z})$. Then $\langle \alpha_i, \alpha_k \rangle = \sum_j m_{ij} \langle \lambda_j, \alpha_k \rangle = m_{ik}$. In other words, the Cartan matrix expresses the change of basis. To write the λ_j in terms of the α_i, we have only to invert the Cartan matrix; its determinant (cf. Exercise 11.2) is the sole denominator involved, so this measures the index of Λ_r in Λ. For example, in type A_1, $\alpha_1 = 2\lambda_1$. (This is the only case in which a simple root is dominant, for reasons which will later become apparent.) In type A_2, the Cartan matrix is $\begin{pmatrix} 2 & -1 \\ -1 & 2 \end{pmatrix}$, so $\alpha_1 = 2\lambda_1 - \lambda_2$ and $\alpha_2 = -\lambda_1 + 2\lambda_2$. Inverting, we get $(1/3)\begin{pmatrix} 2 & 1 \\ 1 & 2 \end{pmatrix}$, so that $\lambda_1 = (1/3)$ $(2\alpha_1 + \alpha_2)$ and $\lambda_2 = (1/3)(\alpha_1 + 2\alpha_2)$. By computing determinants of Cartan matrices one verifies the following list of orders for the fundamental groups Λ/Λ_r in the irreducible cases:

$$A_\ell, \ell+1; \; B_\ell, C_\ell, E_7, 2; \; D_\ell, 4; \; E_6, 3; \; E_8, F_4, G_2, 1.$$

With somewhat more labor one can calculate explicitly the λ_i in terms of the α_j. This information is listed in Table 1, for the reader's convenience, although strictly speaking we shall not need it in what follows. The exact structure of the fundamental group can be found by computing elementary divisors, or can be deduced from Table 1 once the latter is known (Exercise 4).

13.2. Dominant weights

The Weyl group $\mathscr{W}$ of Φ preserves the inner product on E, hence leaves Λ invariant. (In fact, we already made the more precise observation that $\sigma_i \lambda_j = \lambda_j - \delta_{ij}\alpha_i$.) Orbits of weights under $\mathscr{W}$ occur frequently in the study of representations. In view of Lemma 10.3B, we can state the following fact.

Lemma A. *Each weight is conjugate under $\mathscr{W}$ to one and only one dominant weight. If λ is dominant, then $\sigma\lambda \prec \lambda$ for all $\sigma \in \mathscr{W}$, and if λ is strongly dominant, then $\sigma\lambda = \lambda$ only when $\sigma = 1$.* $\square$

As a subset of E, Λ is partially ordered by the relation: $\lambda \succ \mu$ if and only if $\lambda - \mu$ is a sum of positive roots (10.1). Unfortunately, this ordering does not have too close a connection with the property of being dominant; for example, it is easy to have μ dominant, $\mu \prec \lambda$, but λ not dominant (Exercise 2). Our next lemma shows, however, that dominant weights are not *too* badly behaved relative to $\prec$.

Table 1.

A_ℓ: $\lambda_i = \dfrac{1}{\ell+1} [(\ell-i+1)\alpha_1 + 2(\ell-i+1)\alpha_2 + \ldots + (i-1)(\ell-i+1)\alpha_{i-1}$

$\qquad\qquad + i(\ell-i+1)\alpha_i + i(\ell-i)\alpha_{i+1} + \ldots + i\alpha_\ell]$

B_ℓ: $\lambda_i = \alpha_1 + 2\alpha_2 + \ldots + (i-1)\alpha_{i-1} + i(\alpha_i + \alpha_{i+1} + \ldots + \alpha_\ell)$ $\qquad (i < \ell)$

$\qquad \lambda_\ell = \frac{1}{2}(\alpha_1 + 2\alpha_2 + \ldots + \ell\alpha_\ell)$

C_ℓ: $\lambda_i = \alpha_1 + 2\alpha_2 + \ldots + (i-1)\alpha_{i-1} + i(\alpha_i + \ldots + \alpha_{\ell-1} + \frac{1}{2}\alpha_\ell)$

D_ℓ: $\lambda_i = \alpha_1 + 2\alpha_2 + \ldots + (i-1)\alpha_{i-1} + i(\alpha_i + \ldots + \alpha_{\ell-2}) + \frac{1}{2}i(\alpha_{\ell-1} + \alpha_\ell)$ $\qquad (i < \ell-1)$

$\qquad \lambda_{\ell-1} = \frac{1}{2}(\alpha_1 + 2\alpha_2 + \ldots + (\ell-2)\alpha_{\ell-2} + \frac{1}{2}\ell\alpha_{\ell-1} + \frac{1}{2}(\ell-2)\alpha_\ell)$

$\qquad \lambda_\ell = \frac{1}{2}(\alpha_1 + 2\alpha_2 + \ldots + (\ell-2)\alpha_{\ell-2} + \frac{1}{2}(\ell-2)\alpha_{\ell-1} + \frac{1}{2}\ell\alpha_\ell)$

($\Sigma q_i \alpha_i$ is abbreviated $(q_1, \ldots q_\ell)$ in the following lists.)

E_6: $\lambda_1 = \frac{1}{3}(4, 3, 5, 6, 4, 2)$

$\qquad \lambda_2 = (1, 2, 2, 3, 2, 1)$

$\qquad \lambda_3 = \frac{1}{3}(5, 6, 10, 12, 8, 4)$

$\qquad \lambda_4 = (2, 3, 4, 6, 4, 2)$

$\qquad \lambda_5 = \frac{1}{3}(4, 6, 8, 12, 10, 5)$

$\qquad \lambda_6 = \frac{1}{3}(2, 3, 4, 6, 5, 4)$

E_7: $\lambda_1 = (2, 2, 3, 4, 3, 2, 1)$

$\qquad \lambda_2 = \frac{1}{2}(4, 7, 8, 12, 9, 6, 3)$

$\qquad \lambda_3 = (3, 4, 6, 8, 6, 4, 2)$

$\qquad \lambda_4 = (4, 6, 8, 12, 9, 6, 3)$

$\qquad \lambda_5 = \frac{1}{2}(6, 9, 12, 18, 15, 10, 5)$

$\qquad \lambda_6 = (2, 3, 4, 6, 5, 4, 2)$

$\qquad \lambda_7 = \frac{1}{2}(2, 3, 4, 6, 5, 4, 3)$

E_8: $\lambda_1 = (4, 5, 7, 10, 8, 6, 4, 2)$

$\qquad \lambda_2 = (5, 8, 10, 15, 12, 9, 6, 3)$

$\qquad \lambda_3 = (7, 10, 14, 20, 16, 12, 8, 4)$

$\qquad \lambda_4 = (10, 15, 20, 30, 24, 18, 12, 6)$

$\qquad \lambda_5 = (8, 12, 16, 24, 20, 15, 10, 5)$

$\qquad \lambda_6 = (6, 9, 12, 18, 15, 12, 8, 4)$

$\qquad \lambda_7 = (4, 6, 8, 12, 10, 8, 6, 3)$

$\qquad \lambda_8 = (2, 3, 4, 6, 5, 4, 3, 2)$

F_4: $\lambda_1 = (2, 3, 4, 2)$

$\qquad \lambda_2 = (3, 6, 8, 4)$

$\qquad \lambda_3 = (2, 4, 6, 3)$

$\qquad \lambda_4 = (1, 2, 3, 2)$

G_2: $\lambda_1 = (2, 1)$

$\qquad \lambda_2 = (3, 2)$

Lemma B. *Let* $\lambda \in \Lambda^+$. *Then the number of dominant weights* $\mu \prec \lambda$ *is finite.*

Proof. Let $\lambda = \Sigma r_i \alpha_i$, $\mu = \Sigma s_i \alpha_i$. By Exercise 7, r_i, $s_i \in \mathbf{Q}^+$. If $\lambda \succ \mu$, then all $r_i - s_i$ are in $\mathbf{Z}^+$, which allows only finitely many possibilities for the s_i. $\square$

13.3. The weight δ

Recall (Corollary to Lemma 10.2B) that $\delta = \frac{1}{2} \sum_{\alpha > 0} \alpha$, and that $\sigma_i \delta = \delta - \alpha_i$ $(1 \leq i \leq \ell)$. Of course, δ may or may not lie in the root lattice Λ_r (cf. type A_1); but δ does lie in Λ. More precisely:

Lemma A. $\delta = \sum_{j=1}^{\ell} \lambda_j$, *so* δ *is a (strongly) dominant weight.*

Proof. Since $\sigma_i \delta = \delta - \alpha_i$, $(\delta - \alpha_i, \alpha_i) = (\sigma_i^2 \delta, \sigma_i \alpha_i) = (\delta, -\alpha_i)$, or $2(\delta, \alpha_i) = (\alpha_i, \alpha_i)$, or $\langle \delta, \alpha_i \rangle = 1$ $(1 \leq i \leq \ell)$. But $\delta = \sum_i \langle \delta, \alpha_i \rangle \lambda_i$ (cf. (13.1)), so the lemma follows. $\square$

The next lemma is merely an auxiliary result.

Lemma B. *Let* $\mu \in \Lambda^+$, $\nu = \sigma^{-1} \mu$ $(\sigma \in \mathscr{W})$. *Then* $(\nu + \delta, \nu + \delta) \leq (\mu + \delta, \mu + \delta)$, *with equality only if* $\sigma = 1$, *or* $\mu = 0$.

Proof. $(\nu + \delta, \nu + \delta) = (\sigma(\nu + \delta), \sigma(\nu + \delta)) = (\mu + \sigma\delta, \mu + \sigma\delta) = (\mu + \delta, \mu + \delta) - 2(\mu, \delta - \sigma\delta)$. Since μ is dominant, and $\delta - \sigma\delta$ is a sum of positive roots (Lemmas 13.2A and 13.3A), the right side is $\leq (\mu + \delta, \mu + \delta)$, with equality only if $(\mu, \delta - \sigma\delta) = 0$. Unless $\mu = 0$ $(=\nu)$, this forces $\delta - \sigma\delta = 0$, or $\sigma = 1$. (Cf. Lemma 13.2A and Lemma 13.3A). $\square$

13.4. Saturated sets of weights

Certain finite sets of weights, stable under $\mathscr{W}$, play a prominent role in representation theory. We call a subset Π of Λ **saturated** if for all $\lambda \in \Pi$, $\alpha \in \Phi$, and i between 0 and $\langle \lambda, \alpha \rangle$, the weight $\lambda - i\alpha$ also lies in Π. Notice first that any saturated set is automatically stable under $\mathscr{W}$, since $\sigma_\alpha \lambda = \lambda - \langle \lambda, \alpha \rangle \alpha$ and $\mathscr{W}$ is generated by reflections. We say that a saturated set Π has **highest weight** λ $(\lambda \in \Lambda^+)$ if $\lambda \in \Pi$ and $\mu \prec \lambda$ for all $\mu \in \Pi$. *Examples*: (1) The set consisting of 0 alone is saturated, with highest weight 0. (2) The set Φ of all roots of a semisimple Lie algebra, along with 0, is saturated. In case Φ is irreducible, there is a unique highest root (relative to a fixed base Δ of Φ) (Lemma 10.4A), so Π has this root as its highest weight (why?).

Lemma A. *A saturated set of weights having highest weight* λ *must be finite.*

Proof. Use Lemma 13.2B. $\square$

Lemma B. *Let* Π *be saturated, with highest weight* λ. *If* $\mu \in \Lambda^+$ *and* $\mu \prec \lambda$, *then* $\mu \in \Pi$.

Proof. Suppose $\mu' = \mu + \sum_{\alpha \in \Delta} k_\alpha \alpha \in \Pi$ $(k_\alpha \in \mathbf{Z}^+)$. (*Important*: We do not assume that μ' is dominant.) We shall show how to reduce one of the k_α by one while still remaining in Π, thus eventually arriving at the conclusion that $\mu \in \Pi$. Of course, our starting point is the fact that λ itself is such a μ'. Now suppose $\mu' \neq \mu$, so some k_α is positive. From $(\sum_\alpha k_\alpha \alpha, \sum_\alpha k_\alpha \alpha) > 0$, we deduce that $(\sum_\alpha k_\alpha \alpha, \beta) > 0$ for some $\beta \in \Delta$, with $k_\beta > 0$. In particular, $\langle \sum_\alpha k_\alpha \alpha, \beta \rangle$ is positive. Since μ is dominant, $\langle \mu, \beta \rangle$ is nonnegative. There-fore, $\langle \mu', \beta \rangle$ is positive. By definition of saturated set, it is now possible to subtract β once from μ' without leaving Π, thus reducing k_β by one. $\quad\square$

Corollary (of proof). *Let Π be saturated, with highest weight λ. If $\mu \in \Pi$, $\mu \neq \lambda$, then $\mu + \alpha \in \Pi$ for some $\alpha \in \Delta$.*

Proof. If μ is not dominant, then some $\langle \mu, \alpha \rangle$ is strictly negative, so μ, $\mu + \alpha, \ldots, \mu - \langle \mu, \alpha \rangle \alpha$ all lie in Π. If μ is dominant, the proof of the lemma shows that μ is obtained from λ by subtracting one simple root at a time, while remaining in Π. $\quad\square$

From Lemma B emerges a very clear picture of a saturated set Π having highest weight λ: Π consists of all dominant weights lower than or equal to λ in the partial ordering, along with their conjugates under $\mathscr{W}$. In particular, for given $\lambda \in \Lambda^+$, at most one such set Π can exist. Conversely, given $\lambda \in \Lambda^+$, we may simply define Π to be the set consisting of all dominant weights below λ, along with their $\mathscr{W}$-conjugates. Since Π is stable under $\mathscr{W}$, it can be seen to be saturated (Exercise 10), and thanks to Lemma 13.2A, Π has λ as highest weight.

To conclude this section, we prove an inequality which is essential to the application of Freudenthal's formula (§22).

Lemma C. *Let Π be saturated, with highest weight λ. If $\mu \in \Pi$, then $(\mu + \delta, \mu + \delta) \leq (\lambda + \delta, \lambda + \delta)$, with equality only if $\mu = \lambda$.*

Proof. In view of Lemma 13.3B, it is enough to prove this when μ is dominant, $\mu \neq \lambda$. By the Corollary of Lemma B above, $\mu + \alpha \in \Pi$ for some $\alpha \in \Delta$. Write $\mu' = \mu + \alpha$. Then $(\mu' + \delta, \mu' + \delta) - (\mu + \delta, \mu + \delta) = 2(\mu + \delta, \alpha) + (\alpha, \alpha)$. Since $\mu + \delta$ is dominant, the right side is strictly positive: $(\mu + \delta, \mu + \delta) < (\mu' + \delta, \mu' + \delta)$. In turn, we get a further inequality by replacing μ' by the dominant weight ν to which it is conjugate (Lemma 13.3B again). Since Π is finite and the inequalities are strict, we must eventually arrive at a comparison with λ; the lemma follows. $\quad\square$

Exercises

1. Let $\Phi = \Phi_1 \cup \ldots \cup \Phi_t$ be the decomposition of Φ into its irreducible components, with $\Delta = \Delta_1 \cup \ldots \cup \Delta_t$. Prove that Λ decomposes into a direct sum $\Lambda_1 \oplus \ldots \oplus \Lambda_t$; what about Λ^+?
2. Show by example (e.g., for A_2) that $\lambda \notin \Lambda^+$, $\alpha \in \Delta$, $\lambda - \alpha \in \Lambda^+$ is possible.
3. Verify some of the data in Table 1, e.g., for F_4.

4. Using Table 1, show that the fundamental group of A_ℓ is cyclic of order $\ell+1$, while that of D_ℓ is isomorphic to $\mathbf{Z}/4\mathbf{Z}$ (ℓ odd), or $\mathbf{Z}/2\mathbf{Z} \times \mathbf{Z}/2\mathbf{Z}$ (ℓ even). (It is easy to remember which is which, since $A_3 = D_3$.)

5. If Λ' is any subgroup of Λ which includes Λ_r, prove that Λ' is $\mathscr{W}$-invariant. Therefore, we obtain a homomorphism ϕ: Aut $\Phi/\mathscr{W} \to$ Aut (Λ/Λ_r). Prove that ϕ is injective, then deduce that $-1 \in \mathscr{W}$ if and only if $\Lambda_r \supset 2\Lambda$ (cf. Exercise 12.6). Show that $-1 \in \mathscr{W}$ for precisely the irreducible root systems A_1, B_ℓ, C_ℓ, D_ℓ (ℓ even), E_7, E_8, F_4, G_2.

6. Prove that the roots in Φ which are dominant weights are precisely the highest long root and (if two root lengths occur) the highest short root (cf. (10.4) and Exercise 10.11), when Φ is irreducible.

7. If $\varepsilon_1, \ldots, \varepsilon_\ell$ is an *obtuse* basis of the euclidean space E (i.e., all $(\varepsilon_i, \varepsilon_j) \le$ 0 for $i \ne j$), prove that the dual basis is *acute* (i.e., all $(\varepsilon_i^*, \varepsilon_j^*) \ge 0$ for $i \ne j$). [Reduce to the case $\ell = 2$.]

8. Let Φ be irreducible. Without using the data in Table 1, prove that each λ_i is of the form $\sum_j q_{ij}\alpha_j$, where all q_{ij} are *positive* rational numbers. [Deduce from Exercise 7 that all q_{ij} are nonnegative. From $(\lambda_i, \lambda_i) > 0$ obtain $q_{ii} > 0$. Then show that if $q_{ij} > 0$ and $(\alpha_j, \alpha_k) < 0$, then $q_{ik} > 0$.]

9. Let $\lambda \in \Lambda^+$. Prove that $\sigma(\lambda + \delta) - \delta$ is dominant only for $\sigma = 1$.

10. If $\lambda \in \Lambda^+$, prove that the set Π consisting of all dominant weights $\mu \prec \lambda$ and their $\mathscr{W}$-conjugates is saturated, as asserted in (13.4).

11. Prove that each subset of Λ is contained in a unique smallest saturated set, which is finite if the subset in question is finite.

12. For the root system of type A_2, write down the effect of each element of the Weyl group on each of λ_1, λ_2. Using this data, determine which weights belong to the saturated set having highest weight $\lambda_1 + 3\lambda_2$. Do the same for type G_2 and highest weight $\lambda_1 + 2\lambda_2$.

13. Call $\lambda \in \Lambda^+$ **minimal** if $\mu \in \Lambda^+$, $\mu \prec \lambda$ implies that $\mu = \lambda$. Show that each coset of Λ_r in Λ contains precisely one minimal λ. Prove that λ is minimal if and only if the $\mathscr{W}$-orbit of λ is saturated (with highest weight λ), if and only if $\lambda \in \Lambda^+$ and $\langle \lambda, \alpha \rangle = 0, 1, -1$ for *all* roots α. Determine (using Table 1) the nonzero minimal λ for each irreducible Φ, as follows:

$$A_\ell: \lambda_1, \ldots, \lambda_\ell$$
$$B_\ell: \lambda_\ell$$
$$C_\ell: \lambda_1$$
$$D_\ell: \lambda_1, \lambda_{\ell-1}, \lambda_\ell$$
$$E_6: \lambda_1, \lambda_6$$
$$E_7: \lambda_7$$

Notes

Part of the material in this section is drawn from the text and exercises of Bourbaki [2], Chapter VI, §1, No. 9–10 (and Exercise 23). But we have gone somewhat beyond what is usually done outside representation theory in order to emphasize the role played by the root system.

Chapter IV

Isomorphism and Conjugacy Theorems

14. Isomorphism theorem

We return now to the situation of Chapter II: L is a semisimple Lie algebra over the algebraically closed field F of characteristic 0, H is a maximal toral subalgebra of L, $\Phi \subset H^*$ the set of roots of L relative to H. In (8.5) it was shown that the rational span of Φ in H^* is of dimension ℓ over $\mathbf{Q}$, where $\ell = \dim_F H^*$. By extending the base field from $\mathbf{Q}$ to $\mathbf{R}$ we therefore obtain an ℓ-dimensional real vector space E spanned by Φ. Moreover, the symmetric bilinear form dual to the Killing form is carried along to E, making E a euclidean space. Then Theorem 8.5 affirms that Φ is a root system in E.

Our aim in this section is to prove that two semisimple Lie algebras having the same root system are isomorphic. Actually, we can prove a more precise statement, which leads to the construction of certain automorphisms as well.

14.1. Reduction to the simple case

Proposition. *Let L be a simple Lie algebra, H and Φ as above. Then Φ is an irreducible root system in the sense of* (10.4).

Proof. Suppose not. Then Φ decomposes as $\Phi_1 \cup \Phi_2$, where the Φ_i are orthogonal. If $\alpha \in \Phi_1$, $\beta \in \Phi_2$, then $(\alpha+\beta, \alpha) \neq 0$, $(\alpha+\beta, \beta) \neq 0$, so $\alpha+\beta$ cannot be a root, and $[L_\alpha L_\beta] = 0$. This shows that the subalgebra K of L generated by all L_α ($\alpha \in \Phi_1$) is centralized by all L_β ($\beta \in \Phi_2$); in particular, K is a proper subalgebra of L, because $Z(L) = 0$. Furthermore, K is normalized by all L_α ($\alpha \in \Phi_1$), hence by all L_α ($\alpha \in \Phi$), hence by L (Proposition 8.4 (f)). Therefore K is a proper ideal of L, different from 0, contrary to the simplicity of L. ☐

Next let L be an arbitrary semisimple Lie algebra. Then L can be written uniquely as a direct sum $L_1 \oplus \ldots \oplus L_t$ of simple ideals (Theorem 5.2). If H is a maximal toral subalgebra of L, then $H = H_1 \oplus \ldots \oplus H_t$, where $H_i = L_i \cap H$ (cf. Exercise 5.8). Evidently each H_i is a toral algebra in L_i, in fact *maximal toral*: Any toral subalgebra of L_i larger than H_i would automatically be toral in L, centralize all H_j, $j \neq i$, and generate with them a toral subalgebra of L larger than H. Let Φ_i denote the root system of L_i relative to H_i, in the real vector space E_i. If $\alpha \in \Phi_i$, we can just as well view α as a linear function on H, by decreeing that $\alpha(H_j) = 0$ for $j \neq i$. Then α is clearly a root of L relative to H, with $L_\alpha \subset L_i$. Conversely, if $\alpha \in \Phi$, then

$[H_iL_\alpha] \neq 0$ for some i (otherwise H would centralize L_α), and then $L_\alpha \subset L_i$, so $\alpha|_{H_i}$ is a root of L_i relative to H_i. This discussion shows that Φ may be decomposed as $\Phi_1 \cup \ldots \cup \Phi_t$, $\mathsf{E} \cong \mathsf{E}_1 \oplus \ldots \oplus \mathsf{E}_t$ (cf. (11.3)). From the above proposition we obtain:

Corollary. *Let L be a semisimple Lie algebra, with maximal toral subalgebra H and root system Φ. If $L = L_1 \oplus \ldots \oplus L_t$ is the decomposition of L into simple ideals, then $H_i = H \cap L_i$ is a maximal toral subalgebra of L_i, and the corresponding (irreducible) root system Φ_i may be regarded canonically as a subsystem of Φ in such a way that $\Phi = \Phi_1 \cup \ldots \cup \Phi_t$ is the decomposition of Φ into its irreducible components.* $\square$

This corollary reduces the problem of characterizing semisimple Lie algebras by their root systems to the problem of characterizing simple ones by their (irreducible) root systems.

14.2. Isomorphism theorem

First we single out a small set of generators for L.

Proposition. *Let L be a semisimple Lie algebra, H a maximal toral subalgebra of L, Φ the root system of L relative to H. Fix a base Δ of Φ (10.1). Then L is generated (as Lie algebra) by the root spaces L_α, $L_{-\alpha}$ $(\alpha \in \Delta)$; or equivalently, L is generated by arbitrary nonzero root vectors $x_\alpha \in L_\alpha$, $y_\alpha \in L_{-\alpha}$ $(\alpha \in \Delta)$.*

Proof. Let β be an arbitrary positive root (relative to Δ). By the Corollary of Lemma 10.2A, β may be written in the form $\beta = \alpha_1 + \ldots + \alpha_s$, where $\alpha_i \in \Delta$ and where each partial sum $\alpha_1 + \ldots + \alpha_i$ is a root. We know also (Proposition 8.4 (d)) that $[L_\gamma L_\delta] = L_{\gamma+\delta}$ whenever γ, δ, $\gamma + \delta \in \Phi$. Using induction on s, we see easily that L_β lies in the subalgebra of L generated by all L_α $(\alpha \in \Delta)$. Similarly, if β is negative, then L_β lies in the subalgebra of L generated by all $L_{-\alpha}$ $(\alpha \in \Delta)$. But $L = H + \coprod_{\alpha \in \Phi} L_\alpha$, and $H = \sum_{\alpha \in \Phi} [L_\alpha L_{-\alpha}]$, so the proposition follows. $\square$

If $0 \neq x_\alpha \in L_\alpha$ and $0 \neq y_\alpha \in L_{-\alpha}$ $(\alpha \in \Delta)$, with $[x_\alpha y_\alpha] = h_\alpha$, we shall call $\{x_\alpha, y_\alpha\}$ or $\{x_\alpha, y_\alpha, h_\alpha\}$ a **standard set of generators** for L. Recall that h_α is the unique element of $[L_\alpha L_{-\alpha}]$ at which α takes the value 2.

If (L, H) and (L', H') are two pairs, each consisting of a simple Lie algebra and a maximal toral subalgebra, we want to prove that an isomorphism of the corresponding (irreducible) root systems Φ, Φ' will induce an isomorphism of L onto L' sending H onto H'. By definition, an isomorphism $\Phi \to \Phi'$ is induced by an isomorphism $\mathsf{E} \to \mathsf{E}'$ of the ambient euclidean spaces, the latter not necessarily an isometry. However, the root system axioms are unaffected if we multiply the inner product on E or E' by a positive real number. Therefore, *it does no harm to assume that the isomorphism $\Phi \to \Phi'$ comes from an isometry of the euclidean spaces.* Notice next that the isomorphism $\Phi \to \Phi'$ extends uniquely to an isomorphism of vector spaces $\psi: H^* \to H'^*$ (since Φ spans H^* and Φ' spans H'^*). In turn ψ induces an isomorphism $\pi: H \to H'$, via the Killing form identification of H, H' with their duals. Explicitly, if $\alpha \mapsto \alpha'$ denotes the given map $\Phi \mapsto \Phi'$, then $\pi(t_\alpha) = t'_{\alpha'}$, where t_α and $t'_{\alpha'}$ correspond to α, α' (via the Killing form). Since the given isomorphism of Φ and Φ' comes from an isometry between the respective euclidean spaces, we also have $\pi(h_\alpha) = h'_{\alpha'}$, because $h_\alpha = 2t_\alpha/(\alpha, \alpha)$.

Since H, H' are abelian Lie algebras, π can even be regarded as an isomorphism of Lie algebras. What is wanted is a way to extend π to an isomorphism $L \to L'$ (which we shall again denote by π). If such an extension exists, then a moment's thought shows that it *must* send L_α onto $L'_{\alpha'}$, for all $\alpha \in \Phi$. Now the question arises: To what extent can we hope to specify in advance the element of $L'_{\alpha'}$ to which a given $x_\alpha \in L_\alpha$ should be sent? Obviously the choices of the various $x'_{\alpha'}$ ($\alpha' \in \Phi'$) cannot be completely arbitrary: e.g., if we choose x_α, x_β, $x_{\alpha+\beta}$ satisfying $[x_\alpha x_\beta] = x_{\alpha+\beta}$, then we are forced to choose $x'_{\alpha'+\beta'} = [x'_{\alpha'} x'_{\beta'}]$. This line of reasoning suggests that we concentrate on *simple* roots, where the choices can be made independently.

Theorem. *Let L, L' be simple Lie algebras over F, with respective maximal toral subalgebras H, H' and corresponding root systems Φ, Φ'. Suppose there is an isomorphism of Φ onto Φ' (denoted $\alpha \mapsto \alpha'$), inducing $\pi: H \to H'$. Fix a base $\Delta \subset \Phi$, so $\Delta' = \{\alpha' | \alpha \in \Delta\}$ is a base of Φ'. For each $\alpha \in \Delta$, $\alpha' \in \Delta'$, choose arbitrary (nonzero) $x_\alpha \in L_\alpha$, $x'_{\alpha'} \in L'_{\alpha'}$ (i.e., choose an arbitrary Lie algebra isomorphism $\pi_\alpha: L_\alpha \to L'_{\alpha'}$). Then there exists a unique isomorphism $\pi: L \to L'$ extending $\pi: H \to H'$ and extending all the π_α ($\alpha \in \Delta$).*

Proof. The uniqueness of π (if it exists) is immediate: x_α ($\alpha \in \Delta$) determines unique $y_\alpha \in L_{-\alpha}$ for which $[x_\alpha y_\alpha] = h_\alpha$, and L is generated by the x_α, y_α ($\alpha \in \Delta$), by the above proposition.

The idea of the existence proof is not difficult. If L and L' are to be essentially the same, then their direct sum $L \oplus L'$ (a semisimple Lie algebra with unique simple ideals L, L') should include a subalgebra D resembling the "diagonal" subalgebra $\{(x, x) | x \in L\}$ of $L \oplus L$, which is isomorphic to L under the projection of $L \oplus L$ onto either factor. It is easy to construct a suitable subalgebra D of $L \oplus L'$: As above, x_α ($\alpha \in \Delta$) determines unique $y_\alpha \in L_{-\alpha}$ for which $[x_\alpha y_\alpha] = h_\alpha$, and similarly in L'. Let D be generated by the elements $\bar{x}_\alpha = (x_\alpha, x'_{\alpha'})$, $\bar{y}_\alpha = (y_\alpha, y'_{\alpha'})$, $\bar{h}_\alpha = (h_\alpha, h'_{\alpha'})$ for $\alpha \in \Delta$, $\alpha' \in \Delta'$.

The main problem is to show that D is a *proper* subalgebra; conceivably D might contain elements such as $(x_\alpha, x'_{\alpha'})$ and $(x_\alpha, 2x'_{\alpha'})$, where $x_\alpha \in L_\alpha$, $x'_{\alpha'} \in L'_{\alpha'}$ for some roots α, α', in which case D would contain all of L', then all of L, hence all of $L \oplus L'$ (as the reader can easily verify). It is difficult to see directly that such behavior cannot occur, so instead we proceed indirectly.

Because L, L' are simple, Φ and Φ' are irreducible (Proposition 14.1). Therefore Φ, Φ' have unique maximal roots β, β' (relative to Δ, Δ'), which of course correspond under the given isomorphism $\Phi \to \Phi'$ (Lemma A of (10.4)). Choose arbitrary nonzero $x \in L_\beta$, $x' \in L'_{\beta'}$. Set $\bar{x} = (x, x') \in L \oplus L'$, and let M be the subspace of $L \oplus L'$ spanned by all (*) ad $\bar{y}_{\alpha_1}$ ad $\bar{y}_{\alpha_2} \ldots$ ad $\bar{y}_{\alpha_m}$ ($\bar{x}$), where $\alpha_i \in \Delta$ (repetitions allowed). Obviously (*) belongs to $L_{\beta - \Sigma\alpha_i} \oplus$

$L'_{\beta'-\Sigma\alpha'_i}$; in particular, $M \cap (L_\beta \oplus L'_{\beta'})$ is only one dimensional, forcing M to be a *proper* subspace of $L \oplus L'$.

We claim that *our subalgebra D stabilizes M*, which we verify by looking at generators of D. By definition, ad $\bar{y}_\alpha$ stabilizes M ($\alpha \in \Delta$), and by an easy induction based on the fact that $[hy_\alpha]$ is a multiple of y_α, we see that ad $\bar{h}_\alpha$ does likewise. On the other hand, for simple α, we know that ad x_α commutes with all ad y_γ (γ simple) except $\gamma = \alpha$, since $\alpha - \gamma$ is not a root (Lemma 10.1). If we apply ad $\bar{x}_\alpha$ to (*), we can therefore move it past each ad $\bar{y}_\gamma$ except ad $\bar{y}_\alpha$, in which case an extra summand (involving ad $\bar{h}_\alpha$) is introduced. But we have already taken care of this kind of term. Since ad $\bar{x}_\alpha(\bar{x}) = 0$ whenever $\alpha \in \Delta$ ($\alpha + \beta \notin \Phi$, by maximality), we see finally that ad $\bar{x}_\alpha$ stabilizes M.

Now it is clear that D *is a proper subalgebra*: Otherwise M would be a proper nonzero ideal of $L \oplus L'$, but L, L' are the unique ideals of this type (Theorem 5.2), and obviously $M \neq L, M \neq L'$.

We claim that *the projections of D onto the first and second factors of $L \oplus L'$ are (Lie algebra) isomorphisms*. These projections are Lie algebra homomorphisms, by general principles, and they are onto, thanks to the above proposition and the way D was defined. On the other hand, suppose D has nonzero intersection with L (= kernel of projection onto second factor). This means that D contains some $(w, 0)$, $w \neq 0$; so D also contains all (ad $z_{\alpha_1} \ldots$ ad $z_{\alpha_s} (w), 0)$, $\pm\alpha_i \in \Delta$, $z_\alpha = x_\alpha$ or y_α. These elements form a nonzero ideal of L (by the proposition), which must be L itself (L being simple). Thus D includes L. By symmetry, D must also include L', hence all of $L \oplus L'$, which is not the case.

Finally, we observe that the isomorphism $L \to L'$ just obtained via D sends x_α to $x'_{\alpha'}$ ($\alpha \in \Delta$) and h_α to $h'_{\alpha'}$, hence coincides with π on H. This is what was promised. $\square$

The theorem extends easily (Exercise 1) to semisimple algebras. We remark that there is another, higher powered, approach to the proof of the isomorphism theorem, suggested by the above proposition. Namely, write down an explicit *presentation* of L, with generators x_α, y_α, h_α ($\alpha \in \Delta$) and with suitable relations; choose the relations so that all constants involved are dependent solely on the root system Φ. Then any other simple algebra L' having root system isomorphic to Φ will automatically be isomorphic to L. This proof will in fact be given later on (§18), after some preparation; it is less elementary, but has the advantage of leading simultaneously to an existence theorem for semisimple Lie algebras.

14.3. Automorphisms

The isomorphism theorem can be used to good advantage to prove the existence of automorphisms of a semisimple Lie algebra L (with H, Φ as before): Any automorphism of Φ determines an automorphism of H, which can be extended to L. As a useful example, take the map sending each root to its negative. This evidently belongs to Aut Φ (cf. Exercise 12.6), and the induced map on H sends h to $-h$. In particular, if $\sigma: H \to H$ is this isomorphism, $\sigma(h_\alpha) = -h_\alpha$, which by Proposition 8.3(g) is the same as $h_{-\alpha}$. To apply Theorem 14.2, we decree that x_α should be sent to $-y_\alpha$ $(\alpha \in \Delta)$. (Notice that the unique $z \in L_\alpha$ such that $[-y_\alpha z] = h_{-\alpha}$ is just $-x_\alpha$.) According to the theorem, σ extends to an automorphism of L sending x_α to $-y_\alpha$ $(\alpha \in \Delta)$. The preceding parenthetical remark then implies that y_α is sent to $-x_\alpha$ $(\alpha \in \Delta)$. Moreover, σ has order 2, because σ^2 fixes a set of generators of L. To summarize:

Proposition. *L as in Theorem* 14.2 *(but not necessarily simple). Fix (non-zero)* $x_\alpha \in L_\alpha$ $(\alpha \in \Delta)$ *and let* $y_\alpha \in L_{-\alpha}$ *satisfy* $[x_\alpha y_\alpha] = h_\alpha$. *Then there exists an automorphism* σ *of* L, *of order* 2, *satisfying* $\sigma(x_\alpha) = -y_\alpha$, $\sigma(y_\alpha) = -x_\alpha$ $(\alpha \in \Delta)$, $\sigma(h) = -h$ $(h \in H)$. $\square$

For $L = \mathfrak{sl}(2, \mathsf{F})$, the automorphism σ was already discussed in (2.3).

The Weyl group $\mathscr{W}$ of Φ accounts for most of the automorphisms of Φ (12.2). Theorem 14.2 assures the existence of corresponding automorphisms of L, which extend the action of $\mathscr{W}$ on H. If $\sigma \in \mathscr{W}$, it is clear that the extension of σ to an automorphism of L must map L_β to $L_{\sigma^{-1}\beta}$. (Of course, there are various ways of adjusting the scalar multiples involved.) We can also give a direct construction of such an automorphism of L, based on the discussion in (2.3) and independent of Theorem 14.2. It suffices to do this for the reflection σ_α $(\alpha \in \Phi)$. Since ad x_β $(\beta \in \Phi)$ is nilpotent, it makes sense to define the inner automorphism $\tau_\alpha = \exp$ ad $x_\alpha \cdot \exp$ ad $(-y_\alpha) \cdot \exp$ ad x_α. Here $[x_\alpha y_\alpha] = h_\alpha$, as usual. What is the effect of τ_α on H? Write $H = \operatorname{Ker} \alpha \oplus \mathsf{F}h_\alpha$. Clearly, $\tau_\alpha(h) = h$ for all $h \in \operatorname{Ker} \alpha$, while $\tau_\alpha(h_\alpha) = -h_\alpha$ (2.3). Therefore, τ_α and σ_α agree on H. It follows, moreover, that τ_α sends L_β to $L_{\sigma_\alpha \beta}$.

This method of representing reflections (and hence arbitrary elements of $\mathscr{W}$) by elements of Int L has one unavoidable drawback: It does not in general lead to a realization of $\mathscr{W}$ as a subgroup of Int L (cf. Exercise 5).

Exercises

1. Generalize Theorem 14.2 to the case: L semisimple.
2. Let $L = \mathfrak{sl}(2, \mathsf{F})$. If H, H' are any two maximal toral subalgebras of L, prove that there exists an automorphism of L mapping H onto H'.
3. Prove that the subspace M of $L \oplus L'$ introduced in the proof of Theorem 14.2 will actually equal D, if x and x' are chosen carefully.
4. Let σ be as in Proposition 14.3. Is it necessarily true that $\sigma(x_\alpha) = -y_\alpha$ for nonsimple α, where $[x_\alpha y_\alpha] = h_\alpha$?
5. Consider the simple algebra $\mathfrak{sl}(3, \mathsf{F})$ of type A_2. Show that the subgroup of Int L generated by the automorphisms τ_α in (14.3) is strictly larger than the Weyl group (here $\mathscr{S}_3$). [View Int L as a matrix group and compute τ_α^2 explicitly.]
6. Use Theorem 14.2 to construct a subgroup $\Gamma(L)$ of Aut L isomorphic to the group of all graph automorphisms (12.2) of Φ.

7. For each classical algebra (1.2), show how to choose elements $h_\alpha \in H$ corresponding to a base of Φ (cf. Exercise 8.2).

Notes

The proof of Theorem 14.2 is taken from Winter [1]. The automorphism σ discussed in (14.3) will be used in §25 to construct a "Chevalley basis" of L (cf. also Exercise 25.7).

15. Cartan subalgebras

In §14, we proved that a pair (L, H), consisting of a semisimple Lie algebra and a maximal toral subalgebra, is determined up to isomorphism by its root system Φ. However, it is conceivable that another maximal toral subalgebra H' might lead to an entirely different root system Φ'. (This could of course be ruled out in many instances by use of the classification in §11, since dim L = rank Φ + Card Φ. However, types B_ℓ, C_ℓ are indistinguishable from this point of view!)

In order to show that L alone determines Φ, it would surely suffice to prove that all maximal toral subalgebras of L are conjugate under Aut L. This will be done in §16, but in the wider context of an arbitrary Lie algebra L, where the appropriate analogue of H is a "Cartan subalgebra". This wider context actually makes the proof easier, by allowing us to exploit the special properties of solvable Lie algebras. In the present section we prepare the framework; *here F may be of arbitrary characteristic, except where otherwise specified. For technical convenience we still require F to be algebraically closed*, but this could also be weakened: for the main results it is enough that Card F not be "too small" relative to dim L.

15.1. Decomposition of L relative to ad x

Recall from (4.2) that if $t \in \text{End } V$ (V a finite dimensional vector space), then V is the direct sum of all $V_a = \text{Ker } (t - a \cdot 1)^m$, where m is the multiplicity of a as root of the characteristic polynomial of t. Each V_a is invariant under t, and the restriction of t to V_a is the sum of the scalar a and a nilpotent endomorphism.

This applies in particular to the adjoint action of an element x on a Lie algebra L. Write $L = \coprod_{a \in F} L_a(\text{ad } x) = L_0 (\text{ad } x) \oplus L_* (\text{ad } x)$, where $L_* (\text{ad } x)$ denotes the sum of those $L_a(\text{ad } x)$ for which $a \neq 0$. More generally, if K is a subalgebra of L stable under ad x, we may write $K = K_0 (\text{ad } x) \oplus K_* (\text{ad } x)$ even if $x \notin K$.

Lemma. *If $a, b \in F$, then $[L_a (\text{ad } x), L_b (\text{ad } x)] \subset L_{a+b} (\text{ad } x)$. In particular, $L_0 (\text{ad } x)$ is a subalgebra of L, and for $a \neq 0$, each element of $L_a (\text{ad } x)$ is ad-nilpotent.*

Proof. The following formula is a special case of one noted in the proof of Lemma 4.2B:

$$(\operatorname{ad} x - a - b)^m[yz] = \sum_{i=0}^{m} \binom{m}{i} [(\operatorname{ad} x - a)^i(y), (\operatorname{ad} x - b)^{m-i}(z)].$$

It follows that, for sufficiently large m, all terms on the right side are 0, when $y \in L_a$ (ad x), $z \in L_b$ (ad x). $\square$

15.2. Engel subalgebras

According to Lemma 15.1, L_0 (ad x) is a subalgebra of L, for $x \in L$. Following D. W. Barnes, we call it an **Engel subalgebra**. The following two lemmas are basic to our discussion of Cartan subalgebras.

Lemma A. *Let K be a subalgebra of L. Choose $z \in K$ such that L_0 (ad z) is minimal in the collection of all L_0 (ad x), x in K. Suppose that $K \subset L_0$ (ad z). Then L_0 (ad z) $\subset L_0$ (ad x) for all $x \in K$.*

Proof. Begin with fixed, but arbitrary, $x \in K$, and consider the family {ad $(z + cx)|c \in \mathsf{F}$} of endomorphisms of L; since $K_0 = L_0$ (ad z) is a subalgebra of L including K, these endomorphisms stabilize K_0, hence induce endomorphisms of the quotient vector space L/K_0 as well. If T is an indeterminate, we can therefore express the characteristic polynomial of ad $(z + cx)$ as the product $f(T, c)g(T, c)$ of its characteristic polynomials on K_0, L/K_0, respectively. If $r = \dim K_0$, $n = \dim L$, we can write $f(T, c) = T^r + f_1(c)T^{r-1} + \ldots + f_r(c)$, $g(T, c) = T^{n-r} + g_1(c)T^{n-r-1} + \ldots + g_{n-r}(c)$. The reader will see (after translating this into matrix language) that the coefficients $f_i(c)$, $g_i(c)$ are polynomials in c, of degree at most i.

By definition, the eigenvalue 0 of ad z occurs only on the subspace K_0, which means (for the special case $c = 0$) that g_{n-r} is not identically 0 on F. Therefore we can find as many scalars as we please which are not zeros of g_{n-r}; say $c_1, \ldots, c_{r+1}$ are $r+1$ distinct scalars of this sort. To say that $g_{n-r}(c) \neq 0$ is just to say that 0 is not an eigenvalue of ad $(z + cx)$ on the quotient space; this forces all of L_0 (ad $(z + cx)$) to lie in the subspace K_0. But the latter was chosen to be *minimal*, so we conclude that L_0 (ad z) = L_0 (ad $(z + c_ix)$) for $1 \leq i \leq r+1$. This in turn means that ad $(z + c_ix)$ has the sole eigenvalue 0 on L_0 (ad z), i.e., that $f(T, c_i) = T^r$. So each of the polynomials $f_1, \ldots, f_r$ (each of degree at most r) has $r+1$ distinct zeros $c_1, \ldots, c_{r+1}$. This forces each of these polynomials to be identically 0.

We have just shown that L_0 (ad $(z + cx)$) $\supset K_0$ for all $c \in \mathsf{F}$. Since x was arbitrary, we may now replace it by $x - z$, take $c = 1$, and obtain L_0 (ad x) $\supset L_0$ (ad z). $\square$

Lemma B. *If K is a subalgebra of L containing an Engel subalgebra, then $N_L(K) = K$. In particular, Engel subalgebras are self-normalizing.*

Proof. Say $K \supset L_0$ (ad x). Then ad x acts on $N_L(K)/K$ without eigenvalue 0. On the other hand, $x \in K$ implies $[N_L(K)x] \subset K$, so ad x acts trivially on $N_L(K)/K$. Together, these force $K = N_L(K)$. $\square$

15.3. Cartan subalgebras

A **Cartan subalgebra** (abbreviated CSA) of a Lie algebra L is a nilpotent subalgebra which equals its normalizer in L. This definition has the drawback of not implying that CSA's exist (indeed, over finite fields the existence question is not yet fully settled). If L is semisimple (char $F = 0$), then a maximal toral subalgebra H is abelian (hence nilpotent), and $N_L(H) = H$, because $L = H + \coprod_{\alpha \in \Phi} L_\alpha$, with $[H \, L_\alpha] = L_\alpha$ for $\alpha \in \Phi$. So in this case CSA's certainly exist (and play an important role). More generally, we can prove:

Theorem. *Let H be a subalgebra of the Lie algebra L. Then H is a CSA of L if and only if H is a minimal Engel subalgebra (in particular, CSA's exist).*

Proof. First suppose that $H = L_0$ (ad z) is an Engel subalgebra of L; by Lemma B of (15.2), H is self-normalizing. If in addition H properly contains no other Engel subalgebra, then the hypotheses of Lemma A of (15.2) are satisfied (with $H = K$), forcing $H = L_0$ (ad z) $\subset L_0$ (ad x) for all $x \in H$. In particular, $\mathrm{ad}_H \, x$ is nilpotent for $x \in H$. Therefore (Engel's Theorem) H is nilpotent.

Conversely, let H be a CSA of L. Since H is nilpotent, $H \subset L_0$ (ad x) for all $x \in H$. We want equality to hold for at least one x. Suppose, on the contrary, that this never happens. Take L_0 (ad z), $z \in H$, to be as small as possible. Then Lemma A of (15.2) again applies, and we get L_0 (ad x) $\supset L_0$ (ad z) for all $x \in H$. This means that in the representation of H induced on the nonzero vector space L_0 (ad z)$/H$, each $x \in H$ acts as a nilpotent endomorphism. It follows (3.3) that H annihilates some nonzero $y + H$; or, in other words, that there exists $y \notin H$ for which $[Hy] \subset H$. This contradicts the assumption that H is self-normalizing. □

Corollary. *Let L be semisimple (char $F = 0$). Then the CSA's of L are precisely the maximal toral subalgebras of L.*

Proof. We remarked just before the theorem that any maximal toral subalgebra is a CSA. Conversely, let H be a CSA. Observe that if $x = x_s + x_n$ is the Jordan decomposition of x in L, then L_0 (ad x_s) $\subset L_0$ (ad x): any y killed by a power of ad x_s is also killed by a power of ad x, since ad x_n is nilpotent and commutes with ad x_s. Observe also that for $x \in L$ semisimple, L_0 (ad x) $= C_L(x)$, ad x being diagonalizable. Now the CSA H is minimal Engel, of the form L_0 (ad x) (according to the theorem). The above remarks, along with minimality, force $H = L_0$ (ad x_s) $= C_L(x_s)$. But $C_L(x_s)$ evidently includes a maximal toral subalgebra of L, which we already know is a CSA, hence minimal Engel in its own right. We conclude that H is a maximal toral subalgebra. □

As a corollary of this proof, we notice that *a maximal toral subalgebra of a semisimple Lie algebra (char $F = 0$) has the form $C_L(s)$ for some semisimple element s* (cf. Exercise 8.7). Such an element s is called **regular semisimple**.

15.4. *Functorial properties*

Lemma A. *Let* $\phi: L \to L'$ *be an epimorphism of Lie algebras. If H is a* CSA *of L, then* $\phi(H)$ *is a* CSA *of L'.*

Proof. Obviously $\phi(H)$ is nilpotent. Let $A = \text{Ker } \phi$, and identify L' with L/A. If $x + A$ normalizes $H + A$, then $x \in N_L(H + A)$. But $H + A$ includes a CSA (mimimal Engel subalgebra: Theorem 15.3), so the subalgebra $H + A$ is self-normalizing (Lemma B of (15.2)). Therefore, $x \in H + A$, i.e., $\phi(H)$ is self-normalizing. $\square$

Lemma B. *Let* $\phi: L \to L'$ *be an epimorphism of Lie algebras. Let H' be a* CSA *of L',* $K = \phi^{-1}(H')$. *Then any* CSA *H of K is also a* CSA *of L.*

Proof. H is nilpotent, by assumption. By the preceding lemma, $\phi(H)$ is a CSA of $\phi(K) = H'$, forcing $\phi(H) = H'$ (because CSA's are minimal Engel). If $x \in L$ normalizes H, then $\phi(x)$ normalizes $\phi(H)$, whence $\phi(x) \in \phi(H)$, or $x \in H + \text{Ker } \phi$. But $\text{Ker } \phi \subset K$ (by construction), so $x \in H + K \subset K$. Now $x \in N_K(H) = H$, since H is a CSA of K. $\square$

Exercises

1. A semisimple element of $\mathfrak{sl}(n, \mathsf{F})$ is regular if and only if its eigenvalues are all distinct (i.e., if and only if its minimal and characteristic polynomials coincide).
2. Let L be semisimple (char $\mathsf{F} = 0$). Deduce from Exercise 8.7 that the only solvable Engel subalgebras of L are the CSA's.
3. Let L be semisimple (char $\mathsf{F} = 0$), $x \in L$ semisimple. Prove that x is regular if and only if x lies in exactly one CSA.
4. Let H be a CSA of a Lie algebra L. Prove that H is maximal nilpotent, i.e., not properly included in any nilpotent subalgebra of L. Show that the converse is false.
5. Show how to carry out the proof of Lemma A of (15.2) if the field F is only required to be of cardinality exceeding dim L.
6. Exhibit some regular semisimple elements in the classical linear Lie algebras of type B, C, and D (char $\mathsf{F} = 0$).

Notes

The approach to Cartan subalgebras used here is due largely to Barnes [1], who introduced the notion of Engel subalgebra. See also Winter [1].

16. Conjugacy theorems

In this section F *is assumed to be algebraically closed, of characteristic* 0. We are going to prove that, in an arbitrary Lie algebra L over F, all CSA's are conjugate under the group Int L of inner automorphisms (the group generated by all exp ad x, $x \in L$ ad-nilpotent). For L semisimple, this means that all maximal toral subalgebras are conjugate: therefore, L is uniquely determined (up to isomorphism) by its root system relative to any maximal toral subalgebra. As an auxiliary step we shall also prove that all maximal solvable subalgebras of L are conjugate.

16.1. The group $\mathscr{E}(L)$

Let L be a Lie algebra. Call $x \in L$ **strongly ad-nilpotent** if there exists $y \in L$ and some nonzero eigenvalue a of ad y such that $x \in L_a$ (ad y). This forces x to be ad-nilpotent (15.1), so the terminology is reasonable. Denote by $\mathscr{N}(L)$ the set of all strongly ad-nilpotent elements of L, and by $\mathscr{E}(L)$ the subgroup of Int L generated by all exp ad x, $x \in \mathscr{N}(L)$. (Notice that $\mathscr{N}(L)$ is stable under Aut L; therefore, $\mathscr{E}(L)$ is *normal* in Aut L.)

We prefer to work with $\mathscr{E}(L)$ rather than with all of Int L because $\mathscr{E}(L)$ has better functorial properties. (Actually, when L is semisimple, it turns out after the fact that $\mathscr{E}(L) =$ Int L; cf. (16.5).) For example, if K is a sub-algebra of L, then obviously $\mathscr{N}(K) \subset \mathscr{N}(L)$. This permits us to define the subgroup $\mathscr{E}(L; K)$ of $\mathscr{E}(L)$ generated by all exp ad$_L$ x, $x \in \mathscr{N}(K)$. Then $\mathscr{E}(K)$ is obtained simply by taking the restriction of $\mathscr{E}(L; K)$ to K. By contrast, if we take arbitrary $x \in K$ for which ad$_K$ x is nilpotent, we have no control over ad$_L$ x and therefore no such direct relationship between Int K and Int L.

It is clear that, if $\phi: L \to L'$ is an epimorphism, and $y \in L$, then $\phi(L_a(\mathrm{ad}\ y)) = L'_a$ (ad $\phi(y)$). From this we get: $\phi(\mathscr{N}(L)) = \mathscr{N}(L')$.

Lemma. *Let $\phi: L \to L'$ be an epimorphism. If $\sigma' \in \mathscr{E}(L')$, then there exists $\sigma \in \mathscr{E}(L)$ such that the following diagram commutes:*

$$
\begin{array}{ccc}
L & \xrightarrow{\ \phi\ } & L' \\
\sigma \downarrow & & \downarrow \sigma' \\
L & \xrightarrow[\phi]{} & L'
\end{array}
$$

Proof. It suffices to prove this in case $\sigma' = \exp \mathrm{ad}_{L'} x'$, $x' \in \mathscr{N}(L')$. By the preceding remark, $x' = \phi(x)$ for at least one $x \in \mathscr{N}(L)$. For arbitrary $z \in L$, $(\phi \circ \exp \mathrm{ad}_L x)(z) = \phi(z + [xz] + (1/2)[x[xz]] + \ldots) = \phi(z) + [x'\phi(z)] + (1/2)[x'[x'\phi(z)]] + \ldots) = (\exp \mathrm{ad}_{L'} x' \circ \phi)(z)$. In other words, the diagram commutes. $\square$

16.2. Conjugacy of CSA's (solvable case)

Theorem. *Let L be solvable, $\mathscr{E}(L)$ as in (16.1). Then any two CSA's H_1, H_2 of L are conjugate under $\mathscr{E}(L)$.*

Proof. Use induction on dim L, the case dim $L = 1$ (or L nilpotent) being trivial. Assume that L is not nilpotent. Since L is solvable, L possesses nonzero abelian ideals (e.g., the last nonzero term of the derived series); choose A to be one of smallest possible dimension. Set $L' = L/A$, and denote the canonical map $\phi \colon L \to L/A$ by $x \mapsto x'$. According to Lemma A of (15.4), H'_1 and H'_2 are CSA's of the (solvable) algebra L'. By induction, there exists $\sigma' \in \mathscr{E}(L')$ sending H'_1 onto H'_2. Then Lemma 16.1 allows us to find $\sigma \in \mathscr{E}(L)$ such that the diagram there commutes. This means that σ maps the full inverse image $K_1 = \phi^{-1}(H'_1)$ onto $K_2 = \phi^{-1}(H'_2)$. But now H_2 and $\sigma(H_1)$ are both CSA's of the algebra K_2. If K_2 is smaller than L, induction allows us to find $\tau' \in \mathscr{E}(K_2)$ such that $\tau'\sigma(H_1) = H_2$; but $\mathscr{E}(K_2)$ consists of the restrictions to K_2 of the elements of $\mathscr{E}(L; K_2) \subset \mathscr{E}(L)$, so this says that $\tau\sigma(H_1) = H_2$ for $\tau \in \mathscr{E}(L)$ whose restriction to K_2 is τ', and we're done.

Otherwise we must have $L = K_2 = \sigma(K_1)$, so in fact $K_1 = K_2$ and $L = H_2 + A = H_1 + A$. To settle this case, we must explicitly construct an automorphism of L (this is the only point in the whole argument where we have to do so!). The CSA H_2 is of the form L_0 (ad x) for suitable $x \in L$, thanks to Theorem 15.3. A being ad x stable, $A = A_0$ (ad x) $\oplus A_*$ (ad x) (cf. (15.1)), and each summand is stable under $L = H_2 + A$. By the minimality of A, we have either $A = A_0$ (ad x) or else $A = A_*$ (ad x). The first case is absurd, since it would force $A \subset H_2, L = H_2$, contrary to the assumption that L is not nilpotent. So $A = A_*$ (ad x), whence (clearly) $A = L_*$ (ad x).

Since $L = H_1 + A$, we can now express $x = y + z$, where $y \in H_1$, $z \in L_*$ (ad x). In turn, write $z = [xz']$, $z' \in L_*$ (ad x), using the fact that ad x is invertible on L_* (ad x). Since A is abelian, (ad z')$^2 = 0$, so exp ad $z' = 1_L$ $+$ ad z'; applied to x, this yields $x - z = y$. In particular, $H = L_0$ (ad y) must also be a CSA of L. Since $y \in H_1$, $H \supset H_1$, whence $H = H_1$ (both being minimal Engel). So H_1 is conjugate to H_2 via exp ad z'.

It remains only to observe that exp ad z' does lie in $\mathscr{E}(L)$: z' can be written as sum of certain strongly ad-nilpotent elements z_i of $A = L_*$ (ad x), but the latter "commute" (A is abelian), so exp ad $z' = \prod_i$ exp ad $z_i \in \mathscr{E}(L)$. $\square$

16.3. Borel subalgebras

To pass from the solvable case to the general case, we utilize **Borel subalgebras** of a Lie algebra L, which are by definition the maximal solvable subalgebras of L. If we can show that any two Borel subalgebras of L are conjugate under $\mathscr{E}(L)$, then it will follow from Theorem 16.2 that all CSA's of L are conjugate.

Lemma A. *If B is a Borel subalgebra of L, then $B = N_L(B)$.*

Proof. Let x normalize B. Then $B + Fx$ is a subalgebra of L, solvable because $[B + Fx, B + Fx] \subset B$, whence $x \in B$ by maximality of B. $\square$

Lemma B. *If Rad $L \neq L$, then the Borel subalgebras of L are in natural 1–1 correspondence with those of the semisimple Lie algebra $L/$Rad L.*

Proof. Rad L being a solvable ideal of L, $B +$ Rad L is a solvable subalgebra of L for any Borel subalgebra B of L, i.e., Rad $L \subset B$ (by maximality). The lemma follows at once. $\square$

From Lemma B it follows that the essential case is that in which L is *semisimple*. In this situation, let H be a CSA, Φ the root system of L relative to H. Fix a base Δ, and with it a set of positive roots. Set $B(\Delta) = H + \coprod_{\alpha \succ 0} L_\alpha$, $N(\Delta) = \coprod_{\alpha \succ 0} L_\alpha$. Then we know that $B(\Delta)$ is a subalgebra of L, with derived algebra $N(\Delta)$. Furthermore, $N(\Delta)$ *is nilpotent*: If $x \in L_\alpha$ ($\alpha \succ 0$), then application of ad x to root vectors for roots of positive height (relative to Δ) increases height by at least one; this shows how to make the descending central series go to zero. It follows now that $B(\Delta)$ is solvable. In fact, we claim that $B(\Delta)$ *is a Borel subalgebra*: Indeed, let K be any subalgebra of L properly including $B(\Delta)$. Then K, being stable under ad H, must include some L_α for $\alpha \prec 0$. But this forces K to include the simple algebra S_α; in particular, K cannot be solvable.

Lemma C. *Let L be semisimple, with* CSA *H and root system Φ. For each base $\Delta \subset \Phi$, $B(\Delta)$ is a Borel subalgebra of L (called* **standard** *relative to H). All standard Borel subalgebras of L relative to H are conjugate under $\mathscr{E}(L)$.*

Proof. Only the second statement remains to be proved. Recall (14.3) that the reflection σ_α, acting on H, may be extended to an inner automorphism τ_α of L, which is (by construction) in $\mathscr{E}(L)$. It is clear that this automorphism sends $B(\Delta)$ to $B(\sigma_\alpha \Delta)$. Using the fact that the Weyl group is generated by reflections, we see that $\mathscr{E}(L)$ acts transitively on the standard Borels relative to H. $\square$

16.4. Conjugacy of Borel subalgebras

Theorem. *The Borel subalgebras of an arbitrary Lie algebra L are all conjugate under $\mathscr{E}(L)$.*

Corollary. *The Cartan subalgebras of an arbitrary Lie algebra L are conjugate under $\mathscr{E}(L)$.*

Proof of Corollary. Let H, H' be two CSA's of L. Being nilpotent (hence solvable), each lies in at least one Borel subalgebra, say B and B' (respectively). By the theorem, there exists $\sigma \in \mathscr{E}(L)$ such that $\sigma(B) = B'$. Now $\sigma(H)$ and H' are both CSA's of the solvable algebra B', so by Theorem 16.2 there exists $\tau' \in \mathscr{E}(B')$ for which $\tau'\sigma(H) = H'$. But τ' is the restriction to B' of some $\tau \in \mathscr{E}(L; B') \subset \mathscr{E}(L)$ (16.1), so finally $\tau\sigma(H) = H'$, $\tau\sigma \in \mathscr{E}(L)$. $\square$

Proof of Theorem. We proceed by induction on dim L, the case dim $L = 1$ being trivial. By Lemmas 16.1 and 16.3B, along with the induction hypothesis, *we may assume that L is semisimple*. We may further assume that B is a standard Borel subalgebra relative to some CSA. It has to be shown that any other Borel subalgebra B' is conjugate to B under $\mathscr{E}(L)$. If $B \cap B' = B$, there is nothing to prove (since this forces $B' = B$ by maximality). Therefore, we may also use a second (downward) induction on dim $(B \cap B')$, assuming that $B \cap B'$ is properly included in both B and B' (with B standard relative to some CSA).

(1) First suppose that $B \cap B' \neq 0$. Two cases arise:

Case (i): *The set N' of nilpotent elements of $B \cap B'$ is nonzero.* We assert that N' is actually an ideal in $B \cap B'$: It is obviously a subspace, and if $x \in B \cap B'$, $y \in N'$, then $[xy] \in [B \cap B', B \cap B']$, which consists of nilpotent elements according to (4.1), Corollary C. N' is of course not an ideal of L, so its normalizer K is a proper subalgebra of L.

Next we show that $B \cap B'$ is *properly* contained in both $B \cap K$, $B' \cap K$. For consider the action of N' on $B/(B \cap B')$ induced by ad. Each $x \in N'$ acts nilpotently on this vector space, so by Theorem 3.3 there must exist nonzero $y + (B \cap B')$ killed by all $x \in N'$, i.e., such that $[xy] \in B \cap B'$, $y \notin B \cap B'$. But $[xy]$ is also in $[BB]$, so is nilpotent; this forces $[xy] \in N'$, or $y \in N_B(N') = B \cap K$, while $y \notin B \cap B'$. Similarly, $B \cap B'$ is properly contained in $B' \cap K$.

On the other hand, $B \cap K$ and $B' \cap K$ are solvable subalgebras of K. Let C, C' be respective Borel subalgebras of K including them (Figure 1). Since $K \neq L$, induction yields $\sigma \in \mathscr{E}(L; K) \subset \mathscr{E}(L)$ such that $\sigma(C') = C$. Because $B \cap B'$ is a proper (nonzero) subalgebra of both C and C', the second induction hypothesis then yields $\tau \in \mathscr{E}(L)$ such that $\tau\sigma(C') \subset B$ (i.e., τ sends a Borel subalgebra of L including $\sigma(C') = C$ onto B). Finally,

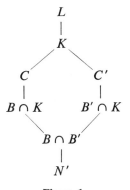

Figure 1

$B \cap \tau\sigma(B') \supset \tau\sigma(C') \cap \tau\sigma(B') \supset \tau\sigma(B' \cap K) \supsetneq \tau\sigma(B \cap B')$, so the former has greater dimension than $B \cap B'$. Again appealing to the second induction hypothesis, we see that B is conjugate under $\mathscr{E}(L)$ to $\tau\sigma(B')$, and we're done with case (i).

Case (ii): *$B \cap B'$ has no nonzero nilpotent elements.* Note that any Borel subalgebra of L contains the semisimple and nilpotent parts of its elements, thanks to Proposition 4.2(c) and Lemma 16.3A. This shows at once that $B \cap B' = T$ is a toral subalgebra. Now we use the fact that B is a standard Borel subalgebra, say $B = B(\Delta)$, $N = N(\Delta)$, $B = H + N$. Since $[BB] = N$, and since $T \cap N = 0$, it is clear that $N_B(T) = C_B(T)$. Let C be any CSA of $C_B(T)$; in particular, C is nilpotent and $T \subset N_{C_B(T)}(C) = C$. The Jacobi identity shows that any element of B which normalizes C also normalizes (hence centralizes) T; therefore, $N_B(C) = N_{C_B(T)}(C) = C$. As a nilpotent self-normalizing subalgebra of B, C is therefore a CSA of B

(which includes T). We know, thanks to Theorem 16.2, that C is a maximal toral subalgebra of L conjugate under $\mathscr{E}(B)$ (hence under $\mathscr{E}(L)$) to H, so without loss of generality *we may now assume that* $T \subset H$.

Suppose $T = H$. Evidently $B' \supsetneq H$, so B' must include at least one L_α ($\alpha \prec 0$ relative to Δ). Applying τ_α (cf. Lemma C of 16.3) to B' yields a Borel subalgebra B'' whose intersection with B includes $H + L_\alpha$; so the second induction hypothesis shows that B'' is conjugate in turn to B, and we're done.

Next suppose T is properly included in H. Now either B' centralizes T or not. If $B' \subset C_L(T)$, then we can appeal to the first induction hypothesis, since dim $C_L(T) <$ dim L ($T \neq 0$ and $Z(L) = 0$). Namely, use the fact that $H \subset C_L(T)$ to find a Borel subalgebra B'' of $C_L(T)$ including H, then use induction to find $\sigma \in \mathscr{E}(L; C_L(T)) \subset \mathscr{E}(L)$ sending B' onto B''. In particular, B'' is a Borel subalgebra of L, including H, so it is conjugate to B under $\mathscr{E}(L)$ because of the second induction hypothesis.

We are left with the situation $B' \not\subset C_L(T)$. This allows us to find a common eigenvector $x \in B'$ for ad T, and an element $t \in T$ for which $[tx] = ax$, with a rational and positive. Define $S = H + \amalg L_\alpha$, $\alpha \in \Phi$ running over those roots for which $\alpha(t)$ is rational and positive. It is clear that S is a subalgebra of L (and $x \in S$). Moreover, it is immediate that S is solvable (cf. proof of Lemma 16.3C). Let B'' be a Borel subalgebra of L which includes S. Now $B'' \cap B' \supset T + \mathsf{F}x \supsetneq T = B' \cap B$, so dim $B'' \cap B' >$ dim $B \cap B'$. Similarly, $B'' \cap B \supset H \supsetneq T$, so dim $B'' \cap B >$ dim $B' \cap B$. The second induction hypothesis, applied to this last inequality, shows that B'' is conjugate to B. (In particular, B'' is obviously standard relative to a CSA conjugate to H.) The second induction hypothesis next applies (because B'' is standard) to the first inequality, showing that B'' is conjugate to B'. So B is conjugate to B'.

(2) This disposes of all cases for which $B \cap B' \neq 0$. Consider now what happens if $B \cap B' = 0$. This forces dim $L \geq$ dim $B +$ dim B'; since B is standard, we know dim $B > (1/2)$ dim L, so B' must be "too small". More precisely, take T to be a maximal toral subalgebra of B'. If $T = 0$, then B' consists of nilpotent elements; B' is therefore nilpotent (Engel's Theorem) as well as self-normalizing (Lemma A of (16.3)), i.e., B' is a CSA. But this is absurd, since we know (Corollary 15.3) that all CSA's of L are toral. Therefore $T \neq 0$. If H_0 is a maximal toral subalgebra of L including T, then B' has nonzero intersection with any standard Borel subalgebra B'' relative to H_0. Therefore B' is conjugate to B'', by the first part of the proof, so dim $B' =$ dim $B'' > (1/2)$ dim L, contradicting the "smallness" of B'. $\square$

Corollary 16.4 allows us to attach a numerical invariant (called **rank**) to an arbitrary Lie algebra L over F, namely, the dimension of a CSA of L. In case L is semisimple, rank L coincides with rank Φ, Φ the root system of L relative to any maximal toral subalgebra ($=$CSA).

It is worthwhile to note one byproduct of the conjugacy theorem for Borel subalgebras. Let L be semisimple, with CSA H and root system Φ. We claim that *any Borel subalgebra B of L which includes H is standard.* Indeed, let $\sigma(B(\Delta)) = B$, where Δ is some given base of Φ, $\sigma \in \mathscr{E}(L)$. Since

H and $\sigma(H)$ are two CSA's of B, they are conjugate under $\mathscr{E}(L; B) \subset \mathscr{E}(L)$ (Theorem 16.2), so we may as well assume that $\sigma(H) = H$. Then it is clear that for each $\alpha \succ 0$, $\sigma(L_\alpha) = L_{\sigma\alpha}$, with $\sigma\alpha$ a root. Moreover, the permutation of roots effected by σ preserves sums, so $\sigma(\Delta) = \Delta'$ is again a base of Φ and $B = B(\Delta')$ is standard.

16.5. Automorphism groups

Let L be semisimple, H a CSA of L, with root system Φ and some fixed base Δ. If τ is any automorphism of L, then of course $\tau(B)$, $B = B(\Delta)$, is another Borel subalgebra of L, so it is sent back to B by some $\sigma_1 \in \mathscr{E}(L)$ (Theorem 16.4). Now H and $\sigma_1\tau(H)$ are two CSA's of L (hence also CSA's of B), so we can find $\sigma_2 \in \mathscr{E}(L; B) \subset \mathscr{E}(L)$ which sends $\sigma_1\tau(H)$ to H (and leaves B invariant), thanks to Theorem 16.2. Since $\sigma_2\sigma_1\tau$ simultaneously preserves H and B, it induces an automorphism of Φ which leaves Δ invariant. From (12.2) we know all such automorphisms: the nontrivial ones arise from nontrivial graph automorphisms, which exist (for Φ irreducible) only in the cases A_ℓ ($\ell > 1$), D_ℓ, E_6. Let ρ be a corresponding automorphism of L (cf. Exercise 14.6). Because ρ is not quite unique, we may adjust the scalars involved in such a way that $\rho\sigma_2\sigma_1\tau$ sends x_α to $c_\alpha x_\alpha$ ($\alpha \succ 0$), y_α to $c_\alpha^{-1} y_\alpha$, hence h_α to h_α (hence all h to themselves). The upshot is that τ differs from an element of the group $\mathscr{E}(L) \cdot \Gamma(L)$, $\Gamma(L) =$ group of **graph automorphisms** of L, only by a **diagonal automorphism**, i.e., an automorphism which is the identity on H and scalar multiplication on each root space L_α.

It can be proved (see Jacobson [1], p. 278) that a diagonal automorphism is always inner; in fact, the construction shows that it can be found in $\mathscr{E}(L)$. Moreover, the product Aut $L = $ Int $(L) \cdot \Gamma(L)$ turns out to be semidirect (Jacobson [1], Chapter IX, exercises), so in particular $\mathscr{E}(L) = $ Int (L). The reader will also find in Jacobson's book detailed descriptions of the automorphism groups for various simple Lie algebras.

Exercises

1. Prove that $\mathscr{E}(L)$ has order one if and only if L is nilpotent.
2. Let L be semisimple, H a CSA, Δ a base of Φ. Prove that any subalgebra of L consisting of nilpotent elements, and maximal with respect to this property, is conjugate under $\mathscr{E}(L)$ to $N(\Delta)$, the derived algebra of $B(\Delta)$.
3. Let Ψ be a set of roots which is **closed** ($\alpha, \beta \in \Psi$, $\alpha + \beta \in \Phi$ implies $\alpha + \beta \in \Psi$) and satisfies $\Psi \cap -\Psi = \varnothing$. Prove that Ψ is included in the set of positive roots relative to some base of Φ. [Use Exercise 2.] (This exercise belongs to the theory of root systems, but is easier to do using Lie algebras.)
4. How does the proof of Theorem 16.4 simplify in case $L = \mathfrak{sl}(2, F)$?
5. Let L be semisimple. Prove that the intersection of two Borel subalgebras of L always includes a CSA of L.
6. Let L be semisimple, $L = H + \amalg L_\alpha$. A subalgebra P of L is called

parabolic if P includes some Borel subalgebra. (In that case P is self-normalizing, by Lemma 15.2B.) Fix a base $\Delta \subset \Phi$, and set $B = B(\Delta)$. For each subset $\Delta' \subset \Delta$, define $P(\Delta')$ to be the subalgebra of L generated by all L_α ($\alpha \in \Delta$ or $-\alpha \in \Delta'$), along with H.

(a) $P(\Delta')$ is a parabolic subalgebra of L (called **standard** relative to Δ).

(b) Each parabolic subalgebra of L including $B(\Delta)$ has the form $P(\Delta')$ for some $\Delta' \subset \Delta$. [Use the Corollary of Lemma 10.2A and Proposition 8.4(d).]

(c) Prove that every parabolic subalgebra of L is conjugate under $\mathscr{E}(L)$ to one of the $P(\Delta')$.

7. Let $L = \mathfrak{sl}(2, \mathsf{F})$, with standard basis (x, h, y). For $c \in \mathsf{F}$, write $x(c) = \exp \operatorname{ad}(cx)$, $y(c) = \exp \operatorname{ad}(cy)$. Define inner automorphisms $w(c) = x(c)y(-c^{-1})x(c)$, $h(c) = w(c)w(1)^{-1}$ ($=w(c)w(-1)$), for $c \neq 0$. Compute the matrices of $w(c)$, $h(c)$ relative to the given basis of L, and deduce that all diagonal automorphisms (16.5) of L are inner. Conclude in this case that $\operatorname{Aut} L = \operatorname{Int} L = \mathscr{E}(L)$.

8. Let L be semisimple. Prove that a semisimple element of L is regular if and only if it lies in only finitely many Borel subalgebras.

Notes

The proof of Theorem 16.4 is due to Winter [1] (inspired in part by G. D. Mostow); see also Barnes [1]. Most of the older proofs use analytic methods ($\mathsf{F} = \mathbf{C}$) or else some algebraic geometry: see Chevalley [2], Jacobson [1], Séminaire "Sophus Lie" [1], Serre [2]. For detailed accounts of the automorphism groups, consult Jacobson [1], Seligman [1].

Chapter V

Existence Theorem

17. Universal enveloping algebras

In this section F *may be an arbitrary field* (*except where otherwise noted*). We shall associate to each Lie algebra L over F an associative algebra with 1 (infinite dimensional, in general), which is generated as "freely" as possible by L subject to the commutation relations in L. This "universal enveloping algebra" is a basic tool in representation theory. Although it could have been introduced right away in Chapter I, we deferred it until now in order to avoid the unpleasant task of proving the Poincaré-Birkhoff-Witt Theorem before it was really needed. The reader is advised to forget temporarily all the specialized theory of semisimple Lie algebras.

17.1. Tensor and symmetric algebras

First we introduce a couple of algebras defined by universal properties. (For further details consult, e.g., S. Lang, Algebra, Reading, Mass.: Addison-Wesley 1965, Ch. XVI.) Fix a finite dimensional vector space V over F. Let $T^0 V = \mathsf{F}$, $T^1 V = V$, $T^2 V = V \otimes V, \ldots, T^m V = V \otimes \ldots \otimes V$ (m copies). Define $\mathfrak{T}(V) = \coprod_{i=0}^{\infty} T^i V$, and introduce an associative product, defined on homogeneous generators of $\mathfrak{T}(V)$ by the obvious rule $(v_1 \otimes \ldots \otimes v_k)$ $(w_1 \otimes \ldots \otimes w_m) = v_1 \otimes \ldots \otimes v_k \otimes w_1 \otimes \ldots \otimes w_m \in T^{k+m} V$. This makes $\mathfrak{T}(V)$ an associative graded algebra with 1, which is generated by 1 along with any basis of V. We call it the **tensor algebra** on V. $\mathfrak{T}(V)$ is the universal associative algebra on n generators ($n = \dim V$), in the following sense: given any F-linear map $\phi \colon V \to \mathfrak{A}$ ($\mathfrak{A}$ an associative algebra with 1 over F), there exists a unique homomorphism of F-algebras $\psi \colon \mathfrak{T}(V) \to \mathfrak{A}$ such that $\psi(1) = 1$ and the following diagram commutes ($i =$ inclusion):

Next let I be the (two sided) ideal in $\mathfrak{T}(V)$ generated by all $x \otimes y - y \otimes x$ ($x, y \in V$) and call $\mathfrak{S}(V) = \mathfrak{T}(V)/I$ the **symmetric algebra** on V; $\sigma \colon \mathfrak{T}(V) \to \mathfrak{S}(V)$ will denote the canonical map. Notice that the generators of I lie in $T^2 V$; this makes it obvious that $I = (I \cap T^2 V) \oplus (I \cap T^3 V) \oplus \ldots$. Therefore, σ is injective on $T^0 V = \mathsf{F}$, $T^1 V = V$ (allowing us to identify V with a subspace of $\mathfrak{S}(V)$), and $\mathfrak{S}(V)$ inherits a grading from $\mathfrak{T}(V)$: $\mathfrak{S}(V)$

$= \coprod_{i=0}^{\infty} S^i V$. The effect of factoring out I is just to make the elements of V commute; so $\mathfrak{S}(V)$ is universal (in the above sense) for linear maps of V into commutative associative F-algebras with 1. Moreover, if $(x_1, \ldots, x_n)$ is any fixed basis of V, then $\mathfrak{S}(V)$ is canonically isomorphic to the polynomial algebra over F in n variables, with basis consisting of 1 and all $x_{i(1)} \cdots x_{i(t)}$, $t \geq 1$, $1 \leq i(1) \leq \ldots \leq i(t) \leq n$.

The reader can easily verify that the preceding constructions go through even when V is infinite dimensional.

For use much later (in §23) we mention a special fact in case char F $= 0$. The symmetric group $\mathscr{S}_m$ acts on $T^m V$ by permuting subscripts of tensors $v_1 \otimes \ldots \otimes v_m$ ($v_i \in V$). An element of $T^m V$ fixed by $\mathscr{S}_m$ is called a **homogeneous symmetric tensor of order m.** *Example*: $x \otimes y + y \otimes x$ (order 2). Fix a basis $(x_1, \ldots, x_n)$ of V, so the products $x_{i(1)} \otimes \ldots \otimes x_{i(m)}$ ($1 \leq i(j) \leq n$) form a basis of $T^m V$. For each ordered sequence $1 \leq i(1) \leq i(2) \ldots \leq i(m) \leq n$, define a symmetric tensor

(*) $$\frac{1}{m!} \sum_{\pi \in \mathscr{S}_m} x_{i(\pi(1))} \otimes \ldots \otimes x_{i(\pi(m))}$$

(which makes sense since $m! \neq 0$ in F). The images of these tensors in $S^m V$ are nonzero and clearly form a basis there, so the tensors (*) in turn must span a complement to $I \cap T^m V$ in $T^m V$. On the other hand, the tensors (*) obviously span the space of all symmetric tensors of order m (call it $\tilde{S}^m V \subset T^m V$). We conclude that σ defines a vector space isomorphism of $\tilde{S}^m V$ onto $S^m V$, hence of the space $\tilde{\mathfrak{S}}(V)$ of all symmetric tensors onto $\mathfrak{S}(V)$.

17.2. Construction of $\mathfrak{U}(L)$

We begin with the abstract definition, for an arbitrary Lie algebra L (allowed here to be infinite dimensional, contrary to our usual convention). A **universal enveloping algebra** of L is a pair $(\mathfrak{U}, i)$, where $\mathfrak{U}$ is an associative algebra with 1 over F, $i: L \to \mathfrak{U}$ is a linear map satisfying

(*) $$i([xy]) = i(x)i(y) - i(y)i(x)$$

for $x, y \in L$, and the following holds: for any associative F-algebra $\mathfrak{A}$ with 1 and any linear map $j: L \to \mathfrak{A}$ satisfying (*), there exists a unique homomorphism of algebras $\phi: \mathfrak{U} \to \mathfrak{A}$ (sending 1 to 1) such that $\phi \circ i = j$.

The *uniqueness* of such a pair $(\mathfrak{U}, i)$ is easy to prove. Given another pair $(\mathfrak{B}, i')$ satisfying the same hypotheses, we get homomorphisms $\phi: \mathfrak{U} \to \mathfrak{B}$, $\psi: \mathfrak{B} \to \mathfrak{U}$. By definition, there is a unique dotted map making the following diagram commute:

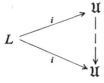

But $1_\mathfrak{U}$ and $\psi \circ \phi$ both do the trick, so $\psi \circ \phi = 1_\mathfrak{U}$. Similarly, $\phi \circ \psi = 1_\mathfrak{B}$.

Existence of a suitable pair ($\mathfrak{U}$, i) is also not difficult to establish. Let $\mathfrak{T}(L)$ be the tensor algebra on L (17.1), and let J be the two sided ideal in $\mathfrak{T}(L)$ generated by all $x \otimes y - y \otimes x - [xy]$ ($x, y \in L$). Define $\mathfrak{U}(L) = \mathfrak{T}(L)/J$, and let $\pi\colon \mathfrak{T}(L) \to \mathfrak{U}(L)$ be the canonical homomorphism. Notice that $J \subset \coprod_{i>0} T^i L$, so π maps $T^0 L = \mathsf{F}$ isomorphically into $\mathfrak{U}(L)$ (therefore, $\mathfrak{U}(L)$ contains at least the scalars). It is not at all obvious that π maps $T^1 L = L$ isomorphically into $\mathfrak{U}(L)$; this will be proved later. In any case, we claim that ($\mathfrak{U}(L)$, i) is a universal enveloping algebra of L, where $i\colon L \to \mathfrak{U}(L)$ is the restriction of π to L. Indeed, let $j\colon L \to \mathfrak{A}$ be as in the definition. The universal property of $\mathfrak{T}(L)$ yields an algebra homomorphism $\phi'\colon \mathfrak{T}(L) \to \mathfrak{A}$ which extends j and sends 1 to 1. The special property (*) of j forces all $x \otimes y - y \otimes x - [xy]$ to lie in Ker ϕ', so ϕ' induces a homomorphism $\phi\colon \mathfrak{U}(L) \to \mathfrak{A}$ such that $\phi \circ i = j$. The uniqueness of ϕ is evident, since 1 and Im i together generate $\mathfrak{U}(L)$.

Example. Let L be *abelian*. Then the ideal J above is generated by all $x \otimes y - y \otimes x$, hence coincides with the ideal I introduced in (17.1). This means that $\mathfrak{U}(L)$ coincides with the symmetric algebra $\mathfrak{S}(L)$. (In particular, $i\colon L \to \mathfrak{U}(L)$ is injective here.)

17.3. PBW *Theorem and consequences*

So far we know very little about the structure of $\mathfrak{U}(L)$, except that it contains the scalars. For brevity, write $\mathfrak{T} = \mathfrak{T}(L)$, $\mathfrak{S} = \mathfrak{S}(L)$, $\mathfrak{U} = \mathfrak{U}(L)$; similarly, write T^m, S^m. Define a *filtration* on $\mathfrak{T}$ by $T_m = T^0 \oplus T^1 \oplus \ldots \oplus T^m$, and let $U_m = \pi(T_m)$, $U_{-1} = 0$. Clearly, $U_m U_p \subset U_{m+p}$ and $U_m \subset U_{m+1}$. Set $G^m = U_m/U_{m-1}$ (this is just a vector space), and let the multiplication in $\mathfrak{U}$ define a bilinear map $G^m \times G^p \to G^{m+p}$. (The map is well-defined; why?) This extends at once to a bilinear map $\mathfrak{S} \times \mathfrak{S} \to \mathfrak{S}$, $\mathfrak{S} = \coprod_{m=0}^{\infty} G^m$, making $\mathfrak{S}$ a graded associative algebra with 1.

Since π maps T^m into U_m, the composite linear map $\phi_m\colon T^m \to U_m \to G^m = U_m/U_{m-1}$ makes sense. It is surjective, because $\pi(T_m - T_{m-1}) = U_m - U_{m-1}$. The maps ϕ_m therefore combine to yield a linear map $\phi\colon \mathfrak{T} \to \mathfrak{S}$, which is surjective (and sends 1 to 1).

Lemma. $\phi\colon \mathfrak{T} \to \mathfrak{S}$ *is an algebra homomorphism. Moreover,* $\phi(I) = 0$, *so ϕ induces a homomorphism ω of* $\mathfrak{S} = \mathfrak{T}/I$ *onto* $\mathfrak{S}$.

Proof. Let $x \in T^m$, $y \in T^p$ be homogeneous tensors. By definition of the product in $\mathfrak{S}$, $\phi(xy) = \phi(x)\phi(y)$, so it follows that ϕ is multiplicative on $\mathfrak{T}$. Let $x \otimes y - y \otimes x$ ($x, y \in L$) be a typical generator of I. Then $\pi(x \otimes y - y \otimes x) \in U_2$, by definition. On the other hand, $\pi(x \otimes y - y \otimes x) = \pi([xy]) \in U_1$, whence $\phi(x \otimes y - y \otimes x) \in U_1/U_1 = 0$. It follows that $I \subset \text{Ker } \phi$. $\square$

The following theorem is the basic result about $\mathfrak{U}(L)$; it (or its Corollary

C) is called the **Poincaré-Birkhoff-Witt Theorem** (or PBW Theorem). The proof will be given in (17.4).

Theorem. *The homomorphism* $\omega\colon \mathfrak{S} \to \mathfrak{G}$ *is an isomorphism of algebras.*

Corollary A. *Let* W *be a subspace of* T^m. *Suppose the canonical map* $T^m \to S^m$ *sends* W *isomorphically onto* S^m. *Then* $\pi(W)$ *is a complement to* U_{m-1} *in* U_m.

Proof. Consider the diagram (all maps canonical):

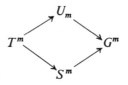

Thanks to the lemma above (and the definitions), this is a commutative diagram. Since $\omega\colon \mathfrak{S} \to \mathfrak{G}$ is an isomorphism (by the theorem), the bottom map sends $W \subset T^m$ isomorphically onto G^m. Reverting to the top map, we get the corollary. ☐

Corollary B. *The canonical map* $i\colon L \to \mathfrak{U}(L)$ *is injective (so* L *may be identified with* $i(L)$).

Proof. This is the special case $W = T^1 \ (=L)$ of Corollary A. ☐

We have allowed L to be infinite dimensional. In practice, the case where L has countable basis is quite adequate for our purposes.

Corollary C. *Let* $(x_1, x_2, x_3, \ldots)$ *be any ordered basis of* L. *Then the elements* $x_{i(1)} \ldots x_{i(m)} = \pi(x_{i(1)} \otimes \ldots \otimes x_{i(m)})$, $m \in \mathbf{Z}^+$, $i(1) \le i(2) \ldots \le i(m)$, *along with* 1, *form a basis of* $\mathfrak{U}(L)$.

Proof. Let W be the subspace of T^m spanned by all $x_{i(1)} \otimes \ldots \otimes x_{i(m)}$, $i(1) \le \ldots \le i(m)$. Evidently W maps isomorphically onto S^m, so Corollary A shows that $\pi(W)$ is a complement to U_{m-1} in U_m. ☐

A basis of $\mathfrak{U}(L)$ of the type just constructed will be referred to simply as a **PBW basis**.

Corollary D. *Let* H *be a subalgebra of* L, *and extend an ordered basis* $(h_1, h_2, \ldots)$ *of* H *to an ordered basis* $(h_1, \ldots, x_1, \ldots)$ *of* L. *Then the homomorphism* $\mathfrak{U}(H) \to \mathfrak{U}(L)$ *induced by the injection* $H \to L \to \mathfrak{U}(L)$ *is itself injective, and* $\mathfrak{U}(L)$ *is a free* $\mathfrak{U}(H)$-*module with free basis consisting of all* $x_{i(1)} \ldots x_{i(m)}$, $i(1) \le i(2) \le \ldots \le i(m)$, *along with* 1.

Proof. These assertions follow at once from Corollary C. ☐

For use much later, we record a special fact.

Corollary E. *Let char* $\mathsf{F} = 0$. *With notation as in* (17.1), *the composite* $S^m L \to \tilde{S}^m L \to U_m$ *of canonical maps is a (linear) isomorphism of* $S^m L$ *onto a complement of* U_{m-1} *in* U_m.

Proof. Use Corollary A, with $W = \tilde{S}^m$. ☐

17.4. *Proof of* PBW *Theorem*

Fix an ordered basis $(x_\lambda; \lambda \in \Omega)$ of L. This choice identifies $\mathfrak{S}$ with the polynomial algebra in indeterminates z_λ ($\lambda \in \Omega$). For each sequence $\Sigma = (\lambda_1, \ldots, \lambda_m)$ of indices (m is called the *length* of Σ), let $z_\Sigma = z_{\lambda_1} \cdots z_{\lambda_m} \in S^m$ and let $x_\Sigma = x_{\lambda_1} \otimes \ldots \otimes x_{\lambda_m} \in T^m$. Call Σ *increasing* if $\lambda_1 \leq \lambda_2 \leq \ldots \leq \lambda_m$, in the given ordering of Ω; by fiat, $\varnothing$ is increasing and $z_\varnothing = 1$. So $\{z_\Sigma | \Sigma$ increasing$\}$ is a basis of $\mathfrak{S}$. Associated with the grading $\mathfrak{S} = \amalg S^m$ is the filtration $S_m = S^0 \oplus \ldots \oplus S^m$. In the following lemmas, write $\lambda \leq \Sigma$ if $\lambda \leq \mu$ for all $\mu \in \Sigma$.

Lemma A. *For each $m \in \mathbf{Z}^+$, there exists a unique linear map $f_m \colon L \otimes S_m \to \mathfrak{S}$ satisfying:*

(A_m) $f_m(x_\lambda \otimes z_\Sigma) = z_\lambda z_\Sigma$ *for* $\lambda \leq \Sigma$, $z_\Sigma \in S_m$.

(B_m) $f_m(x_\lambda \otimes z_\Sigma) - z_\lambda z_\Sigma \in S_k$ *for* $k \leq m$, $z_\Sigma \in S_k$.

(C_m) $f_m(x_\lambda \otimes f_m(x_\mu \otimes z_T)) = f_m(x_\mu \otimes f_m(x_\lambda \otimes z_T)) + f_m([x_\lambda x_\mu] \otimes z_T)$ *for all*
$$z_T \in S_{m-1}.$$

Moreover, the restriction of f_m to $L \otimes S_{m-1}$ agrees with f_{m-1}.

Proof. Notice that the terms in (C_m) all make sense, once (B_m) is proved. Notice too that the restriction of f_m to $L \otimes S_{m-1}$ automatically satisfies (A_{m-1}), (B_{m-1}), (C_{m-1}), so this restricted map must coincide with f_{m-1} because of the asserted uniqueness. To verify existence and uniqueness of f_m, we proceed by induction on m. For $m = 0$, only $z_\Sigma = 1$ occurs; therefore we may let $f_0(x_\lambda \otimes 1) = z_\lambda$ (and extend linearly to $L \otimes S_0$). Evidently (A_0), (B_0), (C_0) are satisfied, and moreover, (A_0) shows that our choice of f_0 is the only possible one.

Assuming the existence of a unique f_{m-1} satisfying (A_{m-1}), (B_{m-1}), (C_{m-1}), we shall show how to extend f_{m-1} to a map f_m. For this it will suffice to define $f_m(x_\lambda \otimes z_\Sigma)$ when Σ is an increasing sequence of length m.

For the case $\lambda \leq \Sigma$, (A_m) cannot hold unless we define $f_m(x_\lambda \otimes z_\Sigma) = z_\lambda z_\Sigma$. In case $\lambda \leq \Sigma$ fails, the first index μ in Σ must be strictly less than λ, so $\Sigma = (\mu, T)$, where of course $\mu \leq T$ and T has length $m-1$. By (A_{m-1}), $z_\Sigma = z_\mu z_T = f_{m-1}(x_\mu \otimes z_T)$. Since $\mu \leq T$, $f_m(x_\mu \otimes z_T) = z_\mu z_T$ is already defined, so the left side of (C_m) becomes $f_m(x_\lambda \otimes z_\Sigma)$. On the other hand, (B_{m-1}) implies that $f_m(x_\lambda \otimes z_T) = f_{m-1}(x_\lambda \otimes z_T) = z_\lambda z_T \pmod{S_{m-1}}$. This shows that the right side of (C_m) is already defined:

$$z_\mu z_\lambda z_T + f_{m-1}(x_\mu \otimes y) + f_{m-1}([x_\lambda x_\mu] \otimes z_T), \, y \in S_{m-1}.$$

The preceding remarks show that f_m can be defined, and in only one way. Moreover, (A_m) and (B_m) clearly hold, as does (C_m) in case $\mu < \lambda$, $\mu \leq T$. But $[x_\mu x_\lambda] = -[x_\lambda x_\mu]$, so (C_m) also holds for $\lambda < \mu$, $\lambda \leq T$. When $\lambda = \mu$, (C_m) is also true. It remains only to consider the case where neither $\lambda \leq T$ nor $\mu \leq T$ is true. Write $T = (\nu, \Psi)$, where $\nu \leq \Psi$, $\nu < \lambda$, $\nu < \mu$. To keep the notation under control, abbreviate $f_m(x \otimes z)$ by xz whenever $x \in L$, $z \in S_m$.

The induction hypothesis insures that $x_\mu z_T = x_\mu(x_\nu z_\Psi) = x_\nu(x_\mu z_\Psi) + [x_\mu x_\nu] z_\Psi$, and $x_\mu z_\Psi = z_\mu z_\Psi + w$ ($w \in S_{m-2}$), by (B_{m-2}). Since $\nu \leq \Psi$, $\nu < \mu$, (C_m) applies already to $x_\lambda(x_\nu(z_\mu z_\Psi))$. By induction, (C_m) also applies to $x_\lambda(x_\nu w)$, therefore to $x_\lambda(x_\nu(x_\mu z_\Psi))$. Consequently: (*) $x_\lambda(x_\mu z_T) = x_\nu(x_\lambda(x_\mu z_\Psi)) + [x_\lambda x_\nu] (x_\mu z_\Psi) + [x_\mu x_\nu] (x_\lambda z_\Psi) + [x_\lambda [x_\mu x_\nu]] z_\Psi$.

Recall that λ, μ are interchangeable throughout this argument. If we interchange them in (*) and subtract the two resulting equations, we get:

$$x_\lambda(x_\mu z_T) - x_\mu(x_\lambda z_T) = x_\nu(x_\lambda(x_\mu z_\Psi)) - x_\mu(x_\lambda z_\Psi) + [x_\lambda [x_\mu x_\nu]] z_\Psi -$$
$$- [x_\mu [x_\lambda x_\nu]] z_\Psi = x_\nu([x_\lambda x_\mu] z_\Psi) + [x_\lambda [x_\mu x_\nu]] z_\Psi$$
$$+ [x_\mu [x_\nu x_\lambda]] z_\Psi = [x_\lambda x_\mu] (x_\nu z_\Psi) + ([x_\nu [x_\lambda x_\mu]]$$
$$+ [x_\lambda [x_\mu x_\nu]] + [x_\mu [x_\nu x_\lambda]]) z_\Psi = [x_\lambda x_\mu] z_T$$

(thanks to the Jacobi identity).

This proves (C_m), and with it the lemma. ☐

Lemma B. *There exists a representation $\rho: L \to \mathfrak{gl}(\mathfrak{S})$ satisfying:*
(a) $\rho(x_\lambda) z_\Sigma = z_\lambda z_\Sigma$ for $\lambda \leq \Sigma$.
(b) $\rho(x_\lambda) z_\Sigma \equiv z_\lambda z_\Sigma \pmod{S_m}$, if Σ has length m.

Proof. Lemma A allows us to define a linear map $f: L \otimes \mathfrak{S} \to \mathfrak{S}$ satisfying (A_m), (B_m), (C_m) for all m (since f_m restricted to $L \otimes S_{m-1}$ is f_{m-1}, by the uniqueness part). In other words $\mathfrak{S}$ becomes an L-module (condition (C_m)), affording a representation ρ which satisfies (a), (b), thanks to (A_m), (B_m). ☐

Lemma C. *Let $t \in T_m \cap J$ ($J = Ker\ \pi$, $\pi: \mathfrak{T} \to \mathfrak{U}$ canonical). Then the homogeneous component t_m of t of degree m lies in I (the kernel of the canonical map $\mathfrak{T} \to \mathfrak{S}$).*

Proof. Write t_m as linear combination of basis elements $x_{\Sigma(i)}$ ($1 \leq i \leq r$), each $\Sigma(i)$ of length m. The Lie homomorphism $\rho: L \to \mathfrak{gl}(\mathfrak{S})$ constructed in Lemma B extends, by the universal property of $\mathfrak{U}$, to an algebra homomorphism (also called ρ) $\mathfrak{T} \to End\ \mathfrak{S}$, with $J \subset Ker\ \rho$. So $\rho(t) = 1$. But $\rho(t).1$ is a polynomial whose term of highest degree is the appropriate combination of the $z_{\Sigma(i)}$ ($1 \leq i \leq r$), by Lemma B. Therefore this combination of the $z_{\Sigma(i)}$ is 0 in $\mathfrak{S}$, and $t_m \in I$ as required. ☐

Proof of **PBW** *Theorem.* Let $t \in T^m$, $\pi: \mathfrak{T} \to \mathfrak{U}$ the canonical map. We must show that $\pi(t) \in U_{m-1}$ implies $t \in I$. But $t \in T^m$, $\pi(t) \in U_{m-1}$ together imply that $\pi(t) = \pi(t')$ for some $t' \in T_{m-1}$, whence $t - t' \in J$. Apply Lemma C to the tensor $t - t' \in T_m \cap J$: the homogeneous component of degree m being t, we get $t \in I$. ☐

17.5. Free Lie algebras

The reader may be familiar with the method of constructing groups by generators and relations. We shall use an analogous method in §18 to construct semisimple Lie algebras. For this one needs the notion of free Lie algebra.

Let L be a Lie algebra over F generated by a set X. We say L is **free** on **X** if, given any mapping ϕ of X into a Lie algebra M, there exists a unique homomorphism $\psi: L \to M$ extending ϕ. The reader can easily verify the *uniqueness* (up to a unique isomorphism) of such an algebra L. As to its *existence*, we begin with a vector space V having X as basis, form the tensor algebra $\mathfrak{T}(V)$ (viewed as Lie algebra via the bracket operation), and let L be the Lie subalgebra of $\mathfrak{T}(V)$ generated by X. Given any map ϕ: $X \to M$, let ϕ be extended first to a linear map $V \to M \subset \mathfrak{U}(M)$, then (canonically) to an associative algebra homomorphism $\mathfrak{T}(V) \to \mathfrak{U}(M)$, or a Lie homomorphism (whose restriction to L is the desired $\psi: L \to M$, since ψ maps the generators X into M).

We remark that if L is free on a set X, then a vector space V can be given an L-module structure simply by assigning to each $x \in X$ an element of the Lie algebra $\mathfrak{gl}(V)$ and extending canonically.

Finally, if L is free on X, and if R is the ideal of L generated by elements f_j (j running over some index set), we call L/R the Lie algebra with **generators** x_i and **relations** $f_j = 0$, where x_i are the images in L/R of the elements of X.

Exercises

1. Prove that if dim $L < \infty$, then $\mathfrak{U}(L)$ has no zero divisors. [Hint: Use the fact that the associated graded algebra $\mathfrak{G}$ is isomorphic to a polynomial algebra.]
2. Let L be the two dimensional nonabelian Lie algebra (1.4), with $[xy] = x$. Prove directly that $i: L \to \mathfrak{U}(L)$ is injective (i.e., that $J \cap L = 0$).
3. If $x \in L$, extend ad x to an endomorphism of $\mathfrak{U}(L)$ by defining ad $x(y) = xy - yx$ ($y \in \mathfrak{U}(L)$). If dim $L < \infty$, prove that each element of $\mathfrak{U}(L)$ lies in a finite dimensional L-submodule. [If x, $x_1, \ldots, x_m \in L$, verify that
$$\text{ad } x(x_1 \ldots x_m) = \sum_{i=1}^{m} x_1 x_2 \ldots \text{ad } x(x_i) \cdots x_m.]$$
4. If L is a free Lie algebra on a set X, prove that $\mathfrak{U}(L)$ is isomorphic to the tensor algebra on a vector space having X as basis.
5. Describe the free Lie algebra on a set $X = \{x\}$.
6. How is the PBW Theorem used in the construction of free Lie algebras?

Notes

Our treatment of the PBW Theorem follows Bourbaki [1]. For another approach, see Jacobson [1].

18. Generators and relations

We can now resume our study of a semisimple Lie algebra L over the algebraically closed field F of characteristic 0. The object is to find a presentation of L by generators and relations which depend only on the root system Φ, thereby proving both the existence and the uniqueness of a semisimple Lie algebra having Φ as root system. In this section, contrary to our general convention, *Lie algebras are allowed to be infinite dimensional*.

18.1. Relations satisfied by L

Let L be a semisimple Lie algebra, H a CSA, Φ the corresponding root system, $\Delta = \{\alpha_1, \ldots, \alpha_\ell\}$ a fixed base. Recall that $\langle \alpha_i, \alpha_j \rangle = \dfrac{2(\alpha_i, \alpha_j)}{(\alpha_j, \alpha_j)} = \alpha_i(h_j)$ $(h_j = h_{\alpha_j})$. Fix a standard set of generators $x_i \in L_{\alpha_i}$, $y_i \in L_{-\alpha_i}$, so that $[x_i y_i] = h_i$.

Proposition. *With the above notation, L is generated by $\{x_i, y_i, h_i | 1 \le i \le \ell\}$, and these generators satisfy at least the following relations*:

(S1) $[h_i h_j] = 0$ $(1 \le i, j \le \ell)$.

(S2) $[x_i y_i] = h_i$, $[x_i y_j] = 0$ if $i \ne j$.

(S3) $[h_i x_j] = \langle \alpha_j, \alpha_i \rangle x_j$, $[h_i y_j] = -\langle \alpha_j, \alpha_i \rangle y_j$.

(S_{ij}^+) $(\operatorname{ad} x_i)^{-\langle \alpha_j, \alpha_i \rangle + 1}(x_j) = 0$ $(i \ne j)$.

(S_{ij}^-) $(\operatorname{ad} y_i)^{-\langle \alpha_j, \alpha_i \rangle + 1}(y_j) = 0$ $(i \ne j)$.

Proof. Proposition 14.2 implies that L is already generated by the x_i and y_i. Relation (S1) is clear, as is (S2), in view of the fact that $\alpha_i - \alpha_j \notin \Phi$ when $i \ne j$ (Lemma 10.1). (S3) is obvious. Consider now (S_{ij}^+) $((S_{ij}^-)$ will follow by symmetry). Since $i \ne j$, $\alpha_j - \alpha_i$ is not a root, and the α_i-string through α_j consists of $\alpha_j, \alpha_j + \alpha_i, \ldots, \alpha_j + q\alpha_i$, where $-q = \langle \alpha_j, \alpha_i \rangle$ (see (9.4) or Proposition 8.4(e)). Since $\operatorname{ad} x_i$ maps x_j successively into the root spaces for $\alpha_j + \alpha_i$, $\alpha_j + 2\alpha_i, \ldots$, (S_{ij}^+) follows. $\square$

Notice that the relations in the proposition involve constants which depend only on the root system. Serre discovered that these form a complete set of defining relations for L (Theorem 18.3 below). As a first step toward proving Serre's Theorem, we shall examine the (possibly infinite dimensional) Lie algebra defined by (S1)–(S3) alone.

18.2. Consequences of (S1)–(S3)

Fix a root system Φ, with base $\Delta = \{\alpha_1, \ldots, \alpha_\ell\}$. Abbreviate the Cartan integer $\langle \alpha_i, \alpha_j \rangle$ by c_{ij}. We begin with a free Lie algebra $\hat{L}$ (see (17.4)) on 3ℓ generators $\{\hat{x}_i, \hat{y}_i, \hat{h}_i | 1 \le i \le \ell\}$. Let $\hat{K}$ be the ideal in $\hat{L}$ generated by the following elements: $[\hat{h}_i \hat{h}_j]$, $[\hat{x}_i \hat{y}_j] - \delta_{ij}\hat{h}_i$, $[\hat{h}_i \hat{x}_j] - c_{ji}\hat{x}_j$, $[\hat{h}_i \hat{y}_j] + c_{ji}\hat{y}_j$. Set $L_o = \hat{L}/\hat{K}$, and let x_i, y_i, h_i be the respective images in L_o of the generators. (In general, $\dim L_o = \infty$.)

The trouble with L_o is that it is defined too abstractly (it might even be trivial, for all we know at this point). To study L_o concretely we attempt to construct a suitable representation of it. The construction which follows is the prototype of one which plays a prominent role in Chapter VI, so the reader is urged to follow the argument closely.

As was noted in (17.5), there is no problem about constructing a module for $\hat{L}$: we need only specify a linear transformation corresponding to each of the 3ℓ generators. Let V be the tensor algebra ($=$ free associative algebra) on a vector space with basis $(v_1, \ldots, v_\ell)$, but forget the product in V. To avoid cumbersome notation we abbreviate $v_{i_1} \otimes \ldots \otimes v_{i_t}$ by $v_{i_1} \ldots v_{i_t}$. These tensors (along with 1) form a basis of V over F. Next, define endomorphisms of V as follows:

$$\begin{cases} \hat{h}_j.1 = 0 \\ \hat{h}_j.v_{i_1} \ldots v_{i_t} = -(c_{i_1 j} + \ldots + c_{i_t j})v_{i_1} \ldots v_{i_t} \end{cases}$$

$$\begin{cases} \hat{y}_j.1 = v_j \\ \hat{y}_j.v_{i_1} \ldots v_{i_t} = v_j v_{i_1} \ldots v_{i_t} \end{cases}$$

$$\begin{cases} \hat{x}_j.1 = 0 = \hat{x}_j.v_i \\ \hat{x}_j.v_{i_1} \ldots v_{i_t} = v_{i_1}(\hat{x}_j.v_{i_2} \ldots v_{i_t}) - \delta_{i_1 j}(c_{i_2 j} + \ldots + c_{i_t j})v_{i_2} \ldots v_{i_t} \end{cases}$$

Then there is a (unique) extension to $\hat{L}$ of this action by its generators, yielding a representation $\hat{\phi} \colon \hat{L} \to \mathfrak{gl}(V)$.

Proposition. *Let* $\hat{K}_o = \mathrm{Ker}\ \hat{\phi}$. *Then* $\hat{K} \subset \hat{K}_o$, *i.e.,* $\hat{\phi}$ *factors through* L_o, *thereby making* V *an* L_o-*module.*

Proof. Notice first that $\hat{h}_j$ acts diagonally on V (relative to the chosen basis of V), so that $\hat{\phi}(\hat{h}_i)$ and $\hat{\phi}(\hat{h}_j)$ commute, i.e., $[\hat{h}_i \hat{h}_j] \in \hat{K}_o$. On the other hand, $\hat{\phi}(\hat{y}_j)$ is simply left multiplication by v_j. (It is only the action of $\hat{x}_j$ which complicates matters.)

Setting $j = i_1$ in the formulas, we obtain: $\hat{x}_i.\hat{y}_j.v_{i_2} \ldots v_{i_t} - \hat{y}_j.\hat{x}_i.v_{i_2} \ldots v_{i_t} = -\delta_{ji}(c_{i_2 i} + \ldots + c_{i_t i})v_{i_2} \ldots v_{i_t} = \delta_{ji}\hat{h}_i.v_{i_2} \ldots v_{i_t}$. Also, $(\hat{x}_i \hat{y}_j - \hat{y}_j \hat{x}_i).1 = 0 = \delta_{ij}\hat{h}_i.1$. Therefore, $[\hat{x}_i \hat{y}_j] - \delta_{ij}\hat{h}_i \in \hat{K}_o$.

Next, $(\hat{h}_i \hat{y}_j - \hat{y}_j \hat{h}_i).1 = \hat{h}_i.v_j = -c_{ji}v_j = -c_{ji}\hat{y}_j.1$. Similarly, $(\hat{h}_i \hat{y}_j - \hat{y}_j \hat{h}_i).v_{i_1} \ldots v_{i_t} = \hat{h}_i.v_j v_{i_1} \ldots v_{i_t} + (c_{i_1 i} + \ldots + c_{i_t i})v_j v_{i_1} \ldots v_{i_t} = -c_{ji}\hat{y}_j.v_{i_1} \ldots v_{i_t}$. Therefore, $[\hat{h}_i \hat{y}_j] + c_{ji}\hat{y}_j \in \hat{K}_o$.

For the remaining step, we make a preliminary observation:

$$(*) \qquad \hat{h}_i.\hat{x}_j.v_{i_1} \ldots v_{i_t} = -(c_{i_1 i} + \ldots + c_{i_t i} - c_{ji})\hat{x}_j.v_{i_1} \ldots v_{i_t}.$$

This is proved by induction on t, starting with the case $t = 0$ (then $v_{i_1} \ldots v_{i_t} = 1$, by convention), where both sides are 0. The induction hypothesis just says that $\hat{x}_j.v_{i_2} \ldots v_{i_t}$ is an eigenvector for $\hat{h}_i$, with eigenvalue $-(c_{i_2 i} + \ldots + c_{i_t i} - c_{ji})$. Multiplying this eigenvector by v_{i_1} on the left evidently produces another eigenvector for $\hat{h}_i$, with eigenvalue $-(c_{i_1 i} + \ldots + c_{i_t i} - c_{ji})$. From these remarks and the definitions $(*)$ follows quickly.

Using $(*)$ we calculate: $(\hat{h}_i \hat{x}_j - \hat{x}_j \hat{h}_i).1 = 0$, $(\hat{h}_i \hat{x}_j - \hat{x}_j \hat{h}_i).v_{i_1} \ldots v_{i_t} = (-(c_{i_1 i} + \ldots + c_{i_t i} - c_{ji}) + (c_{i_1 i} + \ldots + c_{i_t i}))\hat{x}_j.v_{i_1} \ldots v_{i_t} = c_{ji}\hat{x}_j.v_{i_1} \ldots v_{i_t}$. So $[\hat{h}_i \hat{x}_j] - c_{ji}\hat{x}_j \in \hat{K}_o$. Finally, $\hat{K} \subset \hat{K}_o$. □

Theorem. *Given a root system* Φ *with base* $\{\alpha_1, \ldots, \alpha_\ell\}$, *let* L_o *be the Lie* *algebra with generators* $\{x_i, y_i, h_i | 1 \le i \le \ell\}$ *and relations* (S1)–(S3). *Then the* h_i *are a basis for an* ℓ-*dimensional abelian subalgebra* H *of* L_o *and* $L_o = Y + H + X$ (*direct sum of subspaces*), *where* Y (*resp.* X) *is the subalgebra of* L_o *generated by the* y_i (*resp.* x_i).

Proof. This proceeds in steps, using the representation $\phi: L_o \to \mathfrak{gl}(V)$ constructed above: $\phi(x) = \hat{\phi}(\hat{x})$ if x is the image in L_o of $\hat{x} \in \hat{L}$.

(1) $\sum F\hat{h}_j \cap Ker\ \hat{\phi} = 0$. If $\hat{h} = \sum_{j=1}^{\ell} a_j \hat{h}_j$, and $\hat{\phi}(\hat{h}) = 0$, then in particular the eigenvalues $-\sum_j a_j c_{ij}$ $(1 \le i \le \ell)$ of $\hat{\phi}(\hat{h})$ are all 0. But the Cartan matrix (c_{ij}) of Φ is nonsingular, so this forces all $a_j = 0$, i.e., $\hat{h} = 0$.

(2) *The canonical map* $\hat{L} \to L_o$ *sends* $\sum F\hat{h}_j$ *isomorphically onto* $\sum Fh_j$. This follows directly from (1).

(3) *The subspace* $\sum F\hat{x}_j + \sum F\hat{y}_j + \sum F\hat{h}_j$ *of* $\hat{L}$ *maps isomorphically into* L_o. Fix i. The relations (S1)–(S3) include: $[x_i y_i] = h_i$, $[h_i x_i] = 2x_i$, $[h_i y_i] = -2y_i$; so $Fx_i + Fy_i + Fh_i$ is a homomorphic image of $\mathfrak{sl}(2, F)$. But the latter is simple, and $h_i \ne 0$ (step (2)), so $Fx_i + Fy_i + Fh_i$ must be isomorphic to $\mathfrak{sl}(2, F)$. Now the set $\{x_j, y_j, h_j | 1 \le j \le \ell\}$ is linearly independent, because its elements are nonzero and satisfy relations (S1)–(S3) (cf. the eigenvalues of the ad h_j). This proves (3).

(4) $H = \sum Fh_j$ *is an* ℓ-*dimensional abelian subalgebra of* L_o. This follows from (2) and relation (S1).

(5) *If* $[x_{i_1} \ldots x_{i_t}]$ *denotes* $[x_{i_1}[x_{i_2} \ldots [x_{i_{t-1}} x_{i_t}] \ldots] \ldots]$, *then* $[h_j[x_{i_1} \ldots x_{i_t}]] = (c_{i_1 j} + \ldots + c_{i_t j}) [x_{i_1} \ldots x_{i_t}]$, *and similarly for the* y_i *in place of the* x_i, $-c_{ij}$ *in place of* c_{ij}. For $t = 1$ this is (S3). The general case follows quickly by induction, using the Jacobi identity.

(6) *If* $t \ge 2$, *then* $[y_j[x_{i_1} \ldots x_{i_t}]] \in X$, *and similarly for* Y. By relation (S2), $[y_j x_i] = -\delta_{ij} h_i$, and therefore the case $t = 2$ is immediate from the Jacobi identity and (S3). An easy induction on t completes the argument.

(7) $Y + H + X$ *is a subalgebra of* L_o, *hence coincides with* L_o. That $Y + H + X$ is a subalgebra follows from (4), (5), (6). But $Y + H + X$ contains a set of generators of L_o, so it coincides with L_o.

(8) *The sum* $L_o = Y + H + X$ *is direct.* Indeed, (5) shows how to decompose L_o into eigenspaces for ad H; directness follows (cf. (1), (2)). $\square$

It is convenient to describe the decomposition $L_o = Y + H + X$ in terms of "weights" (to use the language of §20). For $\lambda \in H^*$, let $(L_o)_\lambda = \{t \in L_o | [ht] = \lambda(h)t$ for all $h \in H\}$. The proof of the preceding theorem shows that $H = (L_o)_0$. Moreover, the only nonzero λ for which $(L_o)_\lambda \ne 0$ are those of the form $\lambda = \sum_{i=1}^{\ell} k_i \alpha_i$ $(k_i \in \mathbf{Z})$, with all $k_i \ge 0$ (write $\lambda \succ 0$ in this case) or all $k_i \le 0$ (write $\lambda \prec 0$). Then $X = \sum_{\lambda \succ 0} (L_o)_\lambda$ and $Y = \sum_{\lambda \prec 0} (L_o)_\lambda$.

18.3. Serre's Theorem

In (18.2) we studied the structure of the Lie algebra L_o determined by (S1)–(S3) alone. Now we ask what happens when we impose the "finiteness" conditions $(S_{ij}^+), (S_{ij}^-)$ of (18.1). Set $x_{ij} = (\text{ad } x_i)^{-c_{ji}+1}(x_j)$, $y_{ij} = (\text{ad } y_i)^{-c_{ji}+1}(y_j)$ $(i \neq j)$. (These are elements of L_o.)

Lemma. *In the algebra* L_o *of* (18.2), $\text{ad } x_k(y_{ij}) = 0$ $(1 \leq k \leq \ell)$ *for each* $i \neq j$.

Proof. Case (a): $k \neq i$. Then $[x_k y_i] = 0$ (by (S2)), so $\text{ad } x_k$ and $\text{ad } y_i$ commute. Therefore, $\text{ad } x_k(y_{ij}) = (\text{ad } y_i)^{-c_{ji}+1} \text{ad } x_k(y_j)$. If $k = j$, this reads $(\text{ad } y_i)^{-c_{ji}+1}(h_j)$. But by (S3), $\text{ad } y_i(h_j) = c_{ij}y_i$. If this is nonzero, then c_{ji} is nonzero (and negative, since $i \neq j$), so $-c_{ji}+1 \geq 2$. It follows that $(\text{ad } y_i)^{-c_{ji}+1}(h_j) = 0$. If $k \neq j$, then $[x_k y_j] = 0$ (S2), so the same assertion follows.

Case (b): $k = i$. Recall from the proof of Theorem 18.2 that $S = \mathsf{F}x_i + \mathsf{F}y_i + \mathsf{F}h_i$ is a subalgebra of L_o isomorphic to $\mathfrak{sl}(2, \mathsf{F})$. We can therefore say quite a bit about the adjoint action of S on L_o. Even though L_o is (in general) infinite dimensional, some of the reasoning used in §7 carries over directly to the present situation. In particular, since $j \neq i$, $[x_i y_j] = 0$, so that y_j is a "maximal vector" for S, of "weight" $m = -c_{ji}$ (because $[h_i y_j] = -c_{ji}y_j$). An easy induction on t shows that $\text{ad } x_i (\text{ad } y_i)^t(y_j) = t(m-t+1) (\text{ad } y_i)^{t-1}(y_j)$. Therefore the right side is 0 when $t = -c_{ji}+1$. $\square$

Before stating Serre's Theorem, we mention a useful construction. Call an endomorphism x of an infinite dimensional vector space V **locally nilpotent** if every element of V is killed by a sufficiently large power of x. In that case, x is nilpotent on each finite dimensional subspace W of V, so it makes sense to form $\exp(x|_W)$. It is clear that $\exp(x|_W)$ and $\exp(x|_{W'})$ agree on $W \cap W'$, so we can patch these maps together to obtain an automorphism "$\exp x$" of V.

Theorem (Serre). *Fix a root system* Φ, *with base* $\Delta = \{\alpha_1, \ldots, \alpha_\ell\}$. *Let* L *be the Lie algebra generated by* 3ℓ *elements* $\{x_i, y_i, h_i | 1 \leq i \leq \ell\}$, *subject to the relations* (S1), (S2), (S3), (S_{ij}^+), (S_{ij}^-) *listed in* (18.1). *Then* L *is a* (*finite dimensional*) *semisimple algebra, with* CSA *spanned by the* h_i *and with corresponding root system* Φ.

Proof. This will be carried out in steps. By definition, $L = L_o/K$, L_o as in (18.2) and K the ideal generated by all x_{ij}, y_{ij} $(i \neq j)$. To avoid notational problems, we work at first inside L_o. Let I (resp. J) be the ideal of X (resp. Y) generated by all x_{ij} (resp. y_{ij}). (So K includes I, J.)

(1) *I and J are ideals of* L_o. It suffices to consider J (the argument for I being analogous). On the one hand, y_{ij} is an eigenvector for $\text{ad } h_k$ $(1 \leq k \leq \ell)$, with eigenvalue $c_{jk} + (c_{ji}-1)c_{ik}$. Since $\text{ad } h_k(Y) \subset Y$, it follows from the Jacobi identity that $\text{ad } h_k(J) \subset J$. On the other hand, the lemma above says that $\text{ad } x_k(y_{ij}) = 0$. It is clear that $\text{ad } x_k$ maps Y into $Y+H$ (cf. 18.2)); combining this with the Jacobi identity and the fact that $\text{ad } h_k(J) \subset J$, we get $\text{ad } x_k(J) \subset J$. Finally, $\text{ad } L_o(J) \subset J$, again by the Jacobi identity (since the x_k, y_k generate L_o).

(2) $K = I+J$. By definition, $I+J \subset K$. But $I+J$ is an ideal of L_o (by (1)) containing all x_{ij}, y_{ij}, and K is the smallest such ideal.

(3) $L = N^- + H + N$ (*direct sum of subspaces*), *where* $N^- = Y/J$, $N = X/I$, *and* H *is identified with its image under the canonical map* $L_o \to L$. Use (2), along with the direct sum decomposition $L_o = Y + H + X$ (Theorem 18.2).

(4) $\sum Fx_i + \sum Fh_i + \sum Fy_i$ *maps isomorphically into* L. This follows just as in step (3) of the proof of Theorem 18.2, since H maps isomorphically into L (by (3) above). We may therefore identify x_i, y_i, h_i with elements of L (and these generate L).

(5) *If* $\lambda \in H^*$, *let* $L_\lambda = \{x \in L | [hx] = \lambda(h)x \text{ for all } h \in H\}$. *Then* $H = L_0$, $N = \sum_{\lambda > 0} L_\lambda$, $N^- = \sum_{\lambda < 0} L_\lambda$ (*cf. remarks at end of* (18.2)), *and each* L_λ *is finite dimensional*. This is clear, because of (3), (4).

(6) *For* $1 \leq i \leq \ell$, ad x_i *and* ad y_i *are locally nilpotent endomorphisms of* L. It suffices to consider ad x_i, for fixed i (by symmetry). Let M be the subspace of all elements of L which are killed by some power of ad x_i. If $x \in M$ (resp. $y \in M$) is killed by (ad $x_i)^r$ (resp. (ad $x_i)^s$), then $[xy]$ is killed by (ad $x_i)^{r+s}$ (cf. Lemma 15.1). So M is actually a subalgebra of L. But all $x_k \in M$ (by relations (S_{ij}^+)) and all $y_k \in M$ (by (S2), (S3)). These elements generate L, so $M = L$, as desired.

(7) $\tau_i = exp\,(ad\,x_i)\,exp\,(ad\,(-y_i))\,exp\,(ad\,x_i)$ (*for* $1 \leq i \leq \ell$) *is a well-defined automorphism of* L. This follows from (6) and the remarks just preceding the theorem.

(8) *If* λ, $\mu \in H^*$ *and* $\sigma\lambda = \mu$ ($\sigma \in \mathscr{W}$, *the Weyl group of* Φ), *then dim* $L_\lambda = dim\,L_\mu$. It suffices to prove this when $\sigma = \sigma_{\alpha_i}$ is a simple reflection, because these generate $\mathscr{W}$ (Theorem 10.3(d)). The automorphism τ_i of L constructed in step (7) coincides on the finite dimensional space $L_\lambda + L_\mu$ with the ordinary product of exponentials, and we conclude (as in the last part of (7.2)) that τ_i interchanges L_λ, L_μ. In particular, dim $L_\lambda = $ dim L_μ.

(9) *For* $1 \leq i \leq \ell$, *dim* $L_{\alpha_i} = 1$, *while* $L_{k\alpha_i} = 0$ *for integers* $k \neq 0$, 1, -1. This is clear for L_o, hence also for L because of (4).

(10) *If* $\alpha \in \Phi$, *then dim* $L_\alpha = 1$, *but* $L_{k\alpha} = 0$ *for* $k \neq 0$, 1, -1. Each root is $\mathscr{W}$-conjugate to a simple root (Theorem 10.3(c)), so this follows from (8), (9).

(11) *If* $L_\lambda \neq 0$, *then either* $\lambda \in \Phi$ *or* $\lambda = 0$. Otherwise λ is an integral combination of simple roots, with coefficients of like sign (not all 0), λ not a multiple of any root because of (10). Exercise 10.10 shows that some $\mathscr{W}$-conjugate $\sigma\lambda$ has a strictly positive as well as a strictly negative coefficient. This means that $L_{\sigma\lambda} = 0$ (cf. (5)), contradicting the conclusion of step (8).

(12) *dim* $L = \ell + Card\,\Phi < \infty$. In view of (5), this follows from (10), (11).

(13) L *is semisimple*. Let A be an abelian ideal of L; we have to show that $A = 0$. Since ad H stabilizes A, $A = (A \cap H) + \sum_{\alpha \in \Phi}(A \cap L_\alpha)$ (because $L = H + \sum_{\alpha \in \Phi} L_\alpha$). If $L_\alpha \subset A$, then $[L_{-\alpha}L_\alpha] \subset A$, whence $L_{-\alpha} \subset A$ and A contains a copy of the simple algebra $\mathfrak{sl}(2, F)$ (cf. step (4)). This is absurd, so instead $A = A \cap H \subset H$, whence $[L_\alpha A] = 0$ ($\alpha \in \Phi$) and $A \subset \bigcap_{\alpha \in \Phi} \text{Ker } \alpha = 0$ (the α_i span H^*).

(14) H *is a* CSA *of* L, Φ *the root system*. H is abelian (hence nilpotent) and

self-normalizing (because of the direct sum decomposition $L = H + \sum_{\alpha \in \Phi} L_\alpha$),
i.e., H is a CSA. Then Φ is obviously the corresponding set of roots. $\square$

18.4. Application: Existence and uniqueness theorems

Finally, our efforts are rewarded.

Theorem. (a) *Let Φ be a root system. Then there exists a semisimple Lie algebra having Φ as its root system.*

(b) *Let L, L' be semisimple Lie algebras, with respective CSA's H, H' and root systems Φ, Φ'. Let an isomorphism $\Phi \to \Phi'$ be given, sending a given base Δ to a base Δ', and denote by $\pi : H \to H'$ the associated isomorphism (as in (14.2)). For each $\alpha \in \Delta$ ($\alpha' \in \Delta'$) select arbitrary nonzero $x_\alpha \in L_\alpha$ ($x'_{\alpha'} \in L'_{\alpha'}$). Then there exists a unique isomorphism $\pi : L \to L'$ extending $\pi : H \to H'$ and sending x_α to $x'_{\alpha'}$ ($\alpha \in \Delta$).*

Proof. (a) This follows directly from Theorem 18.3. (b) Choose y_α, $y'_{\alpha'}$, (uniquely) satisfying $[x_\alpha y_\alpha] = h_\alpha$, $[x'_{\alpha'} y'_{\alpha'}] = h'_{\alpha'} = \pi(h_\alpha)$ for $\alpha \in \Delta$, $\alpha' \in \Delta'$. Since the $x'_{\alpha'}$, $y'_{\alpha'}$, $h'_{\alpha'}$ ($\alpha' \in \Delta'$) satisfy the relations of Theorem 18.3, which define L, there is a unique homomorphism $\pi : L \to L'$ sending x_α, y_α, h_α ($\alpha \in \Delta$) to $x'_{\alpha'}$, $y'_{\alpha'}$, $h'_{\alpha'}$ (respectively). It is clear that π extends the given isomorphism $H \to H'$. Moreover, the same argument yields a homomorphism $\pi' : L' \to L$, and the composites are the identity maps on generators for L or L', so π is an isomorphism. $\square$

Exercises

1. Using the representation of L_o on V (Proposition 18.2), prove that the algebras X, Y described in Theorem 18.2 are (respectively) free Lie algebras on the sets of x_i, y_i.
2. When rank $\Phi = 1$, the relations (S_{ij}^+), (S_{ij}^-) are vacuous, so $L_o = L \cong \mathfrak{sl}(2, \mathsf{F})$. By suitably modifying the basis of V in (18.2), show that V is isomorphic to the module $Z(0)$ constructed in Exercise 7.7.
3. Prove that the ideal K of L_o in (18.3) lies in every ideal of L_o having finite codimension (i.e., L is the largest finite dimensional quotient of L_o).
4. Prove that each inclusion of Dynkin diagrams (e.g., $\mathsf{E}_6 \subset \mathsf{E}_7 \subset \mathsf{E}_8$) induces a natural inclusion of the corresponding semisimple Lie algebras.

Notes

In essence, the proof of Theorem 18.2 is due (independently) to Chevalley and Harish-Chandra (cf. Harish-Chandra [1]), with simplifications by Jacobson (cf. Jacobson [1]). Serre's Theorem, along with the applications to uniqueness and existence, appears in Serre [2]. See also Varadarajan [1], §5.5.

19. The simple algebras

As in §18, F is algebraically closed of characteristic 0. In this section we assemble some information about the simple Lie algebras over F (much of which has already been indicated in the exercises). According to the classification theorems, there exists one and (up to isomorphism) only one simple Lie algebra having root system A_ℓ ($\ell \geq 1$), B_ℓ ($\ell \geq 2$), C_ℓ ($\ell \geq 3$), D_ℓ ($\ell \geq 4$), E_6, E_7, E_8, F_4, G_2. We shall give rather complete descriptions of the classical types A–D, along with G_2. But for the remaining **exceptional** algebras, the prerequisites concerning Jordan algebras and the like would take us too far afield (see Notes below).

19.1. Criterion for semisimplicity

In theory we can test a given Lie algebra for semisimplicity by computing its Killing form (Theorem 5.1); in practice there is often a much simpler method. First, a definition (cf. Exercise 6.5). A Lie algebra $L \neq 0$ is called **reductive** if Rad $L = Z(L)$. There are two extreme cases: L abelian and L semisimple. $\mathfrak{gl}(V)$ is an intermediate case. Now suppose L is reductive but not abelian, so $L' = L/Z(L)$ is semisimple. Then ad $L \cong$ ad L' acts completely reducibly on L (6.3). Write $L = M \oplus Z(L)$, M an ideal. In particular, $[LL] = [MM] \subset M$. But $[LL]$ maps onto L' under the canonical map, so $L = [LL] \oplus Z(L)$. These remarks imply the first statement of the following proposition.

Proposition. (a) *Let L be reductive. Then $L = [LL] \oplus Z(L)$, and $[LL]$ is either semisimple or 0.*

(b) *Let $L \subset \mathfrak{gl}(V)$ (V finite dimensional) be a nonzero Lie algebra acting irreducibly on V. Then L is reductive, with dim $Z(L) \leq 1$. If in addition $L \subset \mathfrak{sl}(V)$ then L is semisimple.*

Proof. We have to prove (b). Let $S = $ Rad L. By Lie's Theorem, S has a common eigenvector in V, say $s.v = \lambda(s)v$ ($s \in S$). If $x \in L$, then $[sx] \in S$ implies (*) $s.(x.v) = \lambda(s)x.v + \lambda([sx])v$. Since L acts irreducibly, all vectors in V are obtainable by repeated application of elements of L to v and formation of linear combinations. It therefore follows from (*) that the matrices of all $s \in S$ (relative to a suitable basis of V) will be triangular, with $\lambda(s)$ the only diagonal entry. However, the commutators $[SL] \subset S$ have trace 0, so this condition forces λ to vanish on $[SL]$. Referring back to (*), we now conclude that $s \in S$ acts diagonally on V as the scalar $\lambda(s)$. In particular, $S = Z(L)$ (so L is reductive) and dim $S \leq 1$. Finally, let $L \subset \mathfrak{sl}(V)$. Since $\mathfrak{sl}(V)$ contains no scalars except 0 (char F = 0), $S = 0$ and L is semisimple. $\square$

19.2 The classical algebras

In (1.2) we introduced the classical algebras. To avoid repetition, we always limit attention to A_ℓ ($\ell \geq 1$), B_ℓ ($\ell \geq 2$), C_ℓ ($\ell \geq 3$), D_ℓ ($\ell \geq 4$)

(cf. Exercise 1.10 and the classification in §11). Since $\mathfrak{gl}(V) = \mathfrak{sl}(V) +$ (scalars), and since $\mathfrak{gl}(V)$ acts irreducibly on V (it even acts transitively), it is clear that $\mathfrak{sl}(V)$ acts irreducibly as well. This is the prototype of the proof of semisimplicity which we shall give for B_ℓ, C_ℓ, D_ℓ, using the criterion of Proposition 19.1. (We already observed in (1.2) that these algebras consist of endomorphisms of trace 0, so all that needs to be verified is irreducibility.)

Notice that any subspace of V which is invariant under a subalgebra L of $\mathfrak{gl}(V)$ is also invariant under the (associative) subalgebra of End V generated by 1, L. Therefore, to prove that each of B_ℓ, C_ℓ, D_ℓ acts irreducibly in its natural representation, it will suffice to prove that all endomorphisms of V are obtainable from 1 and L using addition, scalar multiplication and ordinary multiplication. From 1 we get all scalars. From the diagonal matrices (as exhibited in (1.2)) we can then get all possible diagonal matrices. Then multiplying various other basis elements (such as $e_{ij} - e_{ji}$, $i \neq j$) by suitable diag $(0, \ldots, 1, \ldots, 0)$ (1 in ith position) yields all the off-diagonal matrix units e_{ij}, as the reader can quickly verify.

The preceding argument shows that the classical algebras are all semisimple. It is clear in each case that the ℓ-dimensional subalgebra H spanned by diagonal matrices (exhibited in (1.2)) is toral and equal to its centralizer in L, hence is a maximal toral subalgebra (= CSA). The remaining basis elements described in (1.2) are root vectors, so it is easy to locate an appropriate set of simple roots, thereby showing that L is simple of the type indicated.

For another approach to the classical algebras, cf. Exercise 6.5.

19.3. The algebra G_2

It will be seen in Chapter VI that the simple algebra of type G_2 has a a (faithful) irreducible representation by 7×7 matrices, and none of smaller degree than 7. It turns out that the representing matrices lie in $L_o = \mathfrak{o}(7, F)$, the simple algebra of type B_3. Since dim $L_o = 21$, while G_2 has dimension 14, it is not too difficult (with the benefit of some hindsight) to describe G_2 directly as a subalgebra L of L_o.

As in (1.2), there is a standard basis for L_o expressed in terms of the matrix units e_{rs} $(1 \leq r, s \leq 7)$. In the following discussion it is convenient to reserve the indices $i, j, k, \ldots$ for the values 1, 2, 3. Recall that L_o has a CSA H_o with basis (d_1, d_2, d_3), $d_i = e_{i+1, i+1} - e_{i+4, i+4}$. Our candidate for CSA of the subalgebra L will be $H = \{\Sigma a_i d_i | \Sigma a_i = 0\}$. Of course, dim $H = 2$.

Corresponding to the six long roots in G_2, which form in their own right a system of type A_2 (Exercise 12.4), we choose certain root vectors $g_{i,-j}$ $(i \neq j)$ of L_o relative to H_o as follows:

$$g_{1,-2} = g_{2,-1}^t = e_{23} - e_{65}.$$

$$g_{1,-3} = g_{3,-1}^t = e_{24} - e_{75}.$$

$$g_{2,-3} = g_{3,-2}^t = e_{34} - e_{76}.$$

For the short roots of L relative to H, we take $g_{\pm i}$ $(i = 1, 2, 3)$:

$$g_1 = -g'_{-1} = \sqrt{2}\,(e_{12}-e_{51})-(e_{37}-e_{46})$$
$$g_2 = -g'_{-2} = \sqrt{2}\,(e_{13}-e_{61})+(e_{27}-e_{45})$$
$$g_3 = -g'_{-3} = \sqrt{2}\,(e_{14}-e_{71})-(e_{26}-e_{35}).$$

Notice that each of the twelve vectors just listed really is a common eigenvector for ad H, with none of them centralizing H. Now we can define L to be the span of H along with these twelve vectors. The following equations imply that L *is closed under the bracket.* The reader can verify them without too much labor (taking advantage of the transpose relationships, to cut down on the number of cases).

(1) $\qquad\qquad [g_{i,-j},\,g_{k,-l}] = \delta_{jk}g_{i,-l}-\delta_{il}g_{k,-j}$

(2) $\qquad\qquad [g_i,\,g_{-i}]\ \ \ = 3d_i-(d_1+d_2+d_3)$

(3) $\qquad\qquad \begin{aligned}[g_{i,-j},\,g_k]\ \ &= -\delta_{ik}g_j \\ [g_{i,-j},\,g_{-k}]\ &= \ \ \delta_{jk}g_{-i}\end{aligned}\Bigg\}$

(4) $\qquad\qquad [g_i,\,g_{-j}]\ \ \ = 3g_{j,-i}$ $\qquad\qquad\qquad (i \neq j)$

(5) $\qquad\qquad \begin{aligned}[g_i,\,g_j]\ \ \ &= \pm 2g_{-k} \\ [g_{-i},\,g_{-j}]\ &= \pm 2g_k\end{aligned}\Bigg\}$ $\qquad (i,\,j,\,k\text{ distinct})$

The signs in (5) can be read off from the equations: $[g_1,\,g_2] = 2g_{-3}$, $[g_1,\,g_3] = -2g_{-2}$, $[g_2,\,g_3] = 2g_{-1}$ (and equations involving transposes).

It follows from what we have said that L is a 14-dimensional Lie algebra, H is a CSA of L (of dimension 2), and that L consists of trace 0 matrices. The classification makes it clear that G_2 is the only possible root system if L is semisimple. Therefore, in view of Proposition 19.1, it remains only to verify that L acts irreducibly on $V = F^7$. Let the canonical ordered basis of V be denoted $(v_0,\,v_1,\,v_2,\,v_3,\,v_{-1},\,v_{-2},\,v_{-3})$. The matrix diag $(0, 1, 2, -3, -1, -2, 3)$ belongs to H and has distinct eigenvalues, so any subspace $W \neq 0$ of V invariant under L must contain at least one of the canonical basis vectors. In turn, observe that $g_{\pm i}$ sends v_0 to a multiple of $v_{\mp i}$, and sends $v_{\pm j}$ to a multiple of v_k $(i, j, k$ distinct), while $g_{i,-j}.v_j = v_i$, $g_{i,-j}.v_{-i} = -v_{-j}$. These equations force W to contain all basis vectors, whence $W = V$ as desired.

It is also interesting to realize the simple algebra of type G_2 as the Lie algebra Der $\mathfrak{C}$ (cf. (1.3)), where $\mathfrak{C}$ is an 8-dimensional nonassociative algebra (the **Cayley** or **octonion algebra**). First it is necessary to describe $\mathfrak{C}$. Let $(e_1,\,e_2,\,e_3)$ be the usual orthonormal basis of F^3 (endowed with its usual inner (or dot) product $v\cdot w$). F^3 also has a vector (or cross) product $v\times w = -(w\times v)$, satisfying the rules: $e_i\times e_i = 0$ $(i = 1, 2, 3)$, $e_1\times e_2 = e_3$, $e_2\times e_3 = e_1$, $e_3\times e_1 = e_2$. As a vector space, $\mathfrak{C}$ is the sum of two copies of F^3 and two copies of F. For convenience, however, we write elements of $\mathfrak{C}$ as 2×2 matrices $\begin{pmatrix} a & v \\ w & b \end{pmatrix}$, where $a,\, b \in F$ and $v,\, w \in F^3$. We add and multiply by scalars just as we would for matrices. However, the product in $\mathfrak{C}$ is given by a more complicated recipe:

$$\begin{pmatrix} a & v \\ w & b \end{pmatrix}\begin{pmatrix} a' & v' \\ w' & b' \end{pmatrix} = \begin{pmatrix} aa' - v \cdot w' & av' + b'v + w \times w' \\ a'w + bw' + v \times v' & bb' - w \cdot v' \end{pmatrix}$$

This operation is obviously bilinear, because the product of scalars in F, and the dot and cross products in F^3, are bilinear.

Fix a basis $(c_1, \ldots, c_8)$ of $\mathfrak{C}$, where

$$c_1 = \begin{pmatrix} 1 & 0 \\ 0 & 0 \end{pmatrix}, \quad c_2 = \begin{pmatrix} 0 & 0 \\ 0 & 1 \end{pmatrix}, \quad c_{2+i} = \begin{pmatrix} 0 & e_i \\ 0 & 0 \end{pmatrix}, \quad c_{5+i} = \begin{pmatrix} 0 & 0 \\ e_i & 0 \end{pmatrix} \quad (i = 1, 2, 3).$$

It is easy to verify the multiplication table for $\mathfrak{C}$ (Table 1). Notice that $c_1 + c_2 = \begin{pmatrix} 1 & 0 \\ 0 & 1 \end{pmatrix}$ acts as identity element for the algebra $\mathfrak{C}$. A routine check of commutators $cc' - c'c$ (using the basis elements and Table 1) shows that

Table 1

	c_1	c_2	c_3	c_4	c_5	c_6	c_7	c_8
c_1	c_1	0	c_3	c_4	c_5	0	0	0
c_2	0	c_1	0	0	0	c_6	c_7	c_8
c_3	0	c_3	0	c_8	$-c_7$	$-c_1$	0	0
c_4	0	c_4	$-c_8$	0	c_6	0	$-c_1$	0
c_5	0	c_5	c_7	$-c_6$	0	0	0	$-c_1$
c_6	c_6	0	$-c_2$	0	0	0	c_5	$-c_4$
c_7	c_7	0	0	$-c_2$	0	$-c_5$	0	c_3
c_8	c_8	0	0	0	$-c_2$	c_4	$-c_3$	0

the subspace $\mathfrak{C}_o$ spanned by all commutators has codimension one, with basis $(c_1 - c_2, c_3, c_4, c_5, c_6, c_7, c_8)$, complementary to the line through $c_1 + c_2$. Moreover, $\mathfrak{C}_o$ then coincides with the space of all elements in $\mathfrak{C}$ having "trace" 0 ($b = -a$). Because of the product rule, any derivation of $\mathfrak{C}$ kills the "constants" (the multiples of $c_1 + c_2$). On the other hand, a derivation obviously leaves $\mathfrak{C}_o$ invariant, hence is completely determined by its restriction to $\mathfrak{C}_o$.

Set $L = \mathrm{Der}\ \mathfrak{C}$. In view of the preceding remarks, L acts faithfully on $\mathfrak{C}_o$ (and trivially on $F(c_1 + c_2)$). Denote by $\phi: L \to \mathfrak{gl}(7, F)$ the associated matrix representation (the basis of $\mathfrak{C}_o$ being chosen as above). The main problem now is to show that L is not too small; for this we actually have to exhibit some derivations of $\mathfrak{C}$. Since the long roots in the root system of type G_2 form a system of type A_2, L ought to include a copy of $\mathfrak{sl}(3, F)$. For $x \in \mathfrak{sl}(3,$ $F)$, define an endomorphism $\delta(x)$ of $\mathfrak{C}$ by $\begin{pmatrix} a & v \\ w & b \end{pmatrix} \mapsto \begin{pmatrix} 0 & x(v) \\ -x^t(w) & 0 \end{pmatrix}$ ($x^t = $ transpose of x). It is a routine matter to verify that $\delta(x)$ is a derivation and that $x \mapsto \delta(x)$ is a (nontrivial, hence faithful) representation of $\mathfrak{sl}(3, F)$. Call the image M, and denote by H the image of the diagonal subalgebra. (Notice that $\phi(M)$ lies in $\mathfrak{o}(7, F)$ and that $\phi(H)$ coincides with the earlier CSA of G_2.)

It is easy to check that the matrices in $\mathfrak{gl}(8, F)$ commuting with H all have the form diag $(a_1, a_2, a_3, x, a_4, a_5, a_6)$, where $x \in \mathfrak{gl}(2, F)$, and that such a matrix represents a derivation of $\mathfrak{C}$ only if it is already in H. Therefore, H *is its own centralizer in* L. Since $Z(M) = 0$, we also deduce that $Z(L) = 0$.

In order to conclude that L is simple of type G_2, we would have to locate more derivations of $\mathfrak{C}$ (corresponding to short roots), then use them to show that L acts irreducibly in $\mathfrak{C}_o$. Actually, we have chosen the basis for $\mathfrak{C}_o$ in such a way that the matrix representation ϕ of L is precisely the one studied above (in $\mathfrak{o}(7, F)$). It would be possible to verify directly that the earlier algebra L consists of the images of derivations of $\mathfrak{C}$ (but quite tedious!). As an alternative approach, certain derivations of $\mathfrak{C}$ (called "inner") can be defined intrinsically and then shown to correspond to the matrices exhibited earlier. This is indicated in the exercises.

Exercises

1. If L is a Lie algebra for which $[LL]$ is semisimple, then L is reductive.
2. Supply details for the argument outlined in (19.2).
3. Let $L \subset \mathfrak{o}(7, F)$ be constructed as in (19.3). Give an alternate proof that L is semisimple, using the fact that L is closed under transposes (cf. Exercise 4.1).
4. Verify the assertions made about $\mathfrak{C}_o$ in (19.3).
5. Verify that $\delta(x)$, $x \in \mathfrak{sl}(3, F)$, as defined in (19.3), is a derivation of $\mathfrak{C}$.
6. Show that the Cayley algebra $\mathfrak{C}$ satisfies the "alternative laws": $x^2 y = x(xy)$, $yx^2 = (yx)x$. Prove that, in any algebra $\mathfrak{A}$ satisfying the alternative laws, an endomorphism of the following form is actually a derivation: $[\lambda_a, \lambda_b] + [\lambda_a, \rho_b] + [\rho_a, \rho_b]$ ($a, b \in \mathfrak{A}$, λ_a = left multiplication in $\mathfrak{A}$ by a, ρ_b = right multiplication in $\mathfrak{A}$ by b, bracket denoting the usual commutator of endomorphisms).
7. Fill in details of the argument at the conclusion of (19.3).

Notes

Tits has constructed the five exceptional simple algebras in a uniform manner; for details and references, see Jacobson [2], Schafer [1]. The characteristic p analogues of the simple Lie algebras discussed here are studied by Seligman [1], cf. Kaplansky [1], Pollack [1]. Our construction of G_2 as a subalgebra of $\mathfrak{o}(7, F)$ is inspired by Exposé 14 of Séminaire "Sophus Lie" [1]. (However, the formulas there contain some errors.)

Chapter VI

Representation Theory

Throughout this chapter L will denote a semisimple Lie algebra (over the algebraically closed field F of characteristic 0), H a fixed CSA of L, Φ the root system, $\Delta = \{\alpha_1, \ldots, \alpha_\ell\}$ a base of Φ, $\mathcal{W}$ the Weyl group. The main object is to study finite dimensional L-modules (although certain infinite dimensional modules will also appear). Thanks to Weyl's Theorem on complete reducibility, it is the irreducible modules which play a controlling role in the finite dimensional case.

20. Weights and maximal vectors

20.1. Weight spaces

If V is a finite dimensional L-module, it follows from Theorem 6.4 that H acts diagonally on V: $V = \coprod V_\lambda$, where λ runs over H^* and $V_\lambda = \{v \in V \mid h.v = \lambda(h)v$ for all $h \in H\}$. For arbitrary V, the subspaces V_λ are still well-defined; whenever $V_\lambda \neq 0$, we call it a **weight space** and we call λ a **weight** of V (more precisely, a "weight of H on V").

Examples. (1) Viewing L itself as an L-module via the adjoint representation, we see that the weights are the roots $\alpha \in \Phi$ (with weight space L_α of dimension one) along with 0 (with weight space H of dimension ℓ). (2) When $L = \mathfrak{sl}(2, \text{F})$ a linear function λ on H is completely determined by its value $\lambda(h)$ at the basis vector h; so we were in effect utilizing weights in §7. (The reader is urged to review that section now.)

If $\dim V = \infty$, there is no assurance that V is the sum of its weight spaces (Exercise 2). Nevertheless, the sum V' of all weight spaces V_λ is always direct: this is essentially the same argument as the one proving that eigenvectors of distinct eigenvalues for a single linear transformation are linearly independent (Exercise 1). Moreover, V' is an L-submodule of V: this follows from the fact that L_α $(\alpha \in \Phi)$ permutes the weight spaces. Namely, if $x \in L_\alpha$, $v \in V_\lambda$, $h \in H$, then $h.x.v = x.h.v + [hx].v = (\lambda(h) + \alpha(h))x.v$, so L_α sends V_λ into $V_{\lambda+\alpha}$. Summarizing:

Lemma. *Let V be an arbitrary L-module. Then*

(a) L_α *maps* V_λ *into* $V_{\lambda+\alpha}$ $(\lambda \in H^*, \alpha \in \Phi)$.

(b) *The sum* $V' = \sum\limits_{\lambda \in H^*} V_\lambda$ *is direct, and* V' *is an L-submodule of V.*

(c) *If $\dim V < \infty$, then $V = V'$.* $\square$

107

20.2. Standard cyclic modules

By definition, a **maximal vector** (of weight λ) in an L-module V is a non-zero vector $v^+ \in V_\lambda$ killed by all L_α ($\alpha \succ 0$, or just $\alpha \in \Delta$). This notion of course depends on the choice of Δ. For example, if L is simple, and β is the maximal root in Φ relative to Δ (Lemma 10.4A), then any nonzero element of L_β is a maximal vector for the adjoint representation of L; these are obviously the only possible maximal vectors in this case. When $\dim V = \infty$, there is no need for a maximal vector to exist. By contrast, if $\dim V < \infty$, then the *Borel subalgebra* (16.3) $B(\Delta) = H + \coprod_{\alpha \succ 0} L_\alpha$ has a common eigenvector (killed by all L_α, $\alpha \succ 0$), thanks to Lie's Theorem, and this is a maximal vector in the above sense.

In order to study finite dimensional irreducible L-modules, it is useful to study first the larger class of L-modules generated by a maximal vector. If $V = \mathfrak{U}(L).v^+$ for a maximal vector v^+ (of weight λ), we say briefly that V is **standard cyclic** (of weight λ) and we call λ the **highest weight** of V. It is easy to describe the structure of such a module. Fix nonzero $x_\alpha \in L_\alpha$ ($\alpha \succ 0$), and choose $y_\alpha \in L_{-\alpha}$ (uniquely) for which $[x_\alpha y_\alpha] = h_\alpha$. Recall the partial ordering $\lambda \succ \mu$ iff $\lambda - \mu$ is a sum of positive roots ($\lambda, \mu \in H^*$), introduced in §10 for the euclidean space E but equally definable for H^*. Part (b) of the following theorem justifies the terminology "highest weight" for λ.

Theorem. *Let V be a standard cyclic L-module, with maximal vector $v^+ \in V_\lambda$. Let $\Phi^+ = \{\beta_1, \ldots, \beta_m\}$. Then:*

(a) V is spanned by the vectors $y_{\beta_1}^{i_1} \ldots y_{\beta_m}^{i_m}.v^+$ ($i_j \in \mathbf{Z}^+$); in particular, V is the direct sum of its weight spaces.

(b) The weights of V are of the form $\mu = \lambda - \sum_{i=1}^{\ell} k_i \alpha_i$ ($k_i \in \mathbf{Z}^+$), i.e., all weights satisfy $\mu \prec \lambda$.

(c) For each $\mu \in H^*$, $\dim V_\mu < \infty$, and $\dim V_\lambda = 1$.

(d) V is an indecomposable L-module, with a unique maximal (proper) submodule.

(e) Every nonzero homomorphic image of V is also standard cyclic of weight λ.

Proof. $L = N^- + B$, where $N^- = \coprod_{\alpha \prec 0} L_\alpha$ and $B = B(\Delta)$. From the PBW Theorem (Corollaries C, D of Theorem 17.3) it follows that $\mathfrak{U}(L).v^+ = \mathfrak{U}(N^-)\mathfrak{U}(B).v^+ = \mathfrak{U}(N^-).Fv^+$ (since v^+ is a common eigenvector for B). Now $\mathfrak{U}(N^-)$ has a basis consisting of monomials $y_{\beta_1}^{i_1} \ldots y_{\beta_m}^{i_m}$, so (a) follows.

The vector (*) $y_{\beta_1}^{i_1} \ldots y_{\beta_m}^{i_m}.v^+$ has weight $\lambda - \sum_j i_j \beta_j$ (Lemma 20.1(a)). Rewriting each β_j as a nonnegative $\mathbf{Z}$-linear combination of simple roots (as in §10), we get (b).

Evidently there are only a finite number of the vectors (*) in (b) for which $\sum i_j \beta_j$ equals a prescribed $\sum_{i=1}^{\ell} k_i \alpha_i$. In view of (a), these span the weight space V_μ, if $\mu = \lambda - \sum k_i \alpha_i$. Moreover, the only vector of the form (*) which has weight $\mu = \lambda$ is v^+ itself, whence (c).

As to (d), suppose that $V = V_1 \oplus V_2$ (direct sum of L-submodules). Write $v^+ = v_1 + v_2$ accordingly. Then for all $b \in B$, $b.v^+ = b.v_1 + b.v_2$ is proportional to v^+, forcing each of v_1, v_2 to be a maximal vector of weight λ (if $v_i \neq 0$). By part (c), $v_i \neq 0$ therefore implies that v_i is a multiple of v^+. The directness of the sum forces either $v_1 = 0$ or else $v_2 = 0$, so $v^+ \in V_2$ or else $v^+ \in V_1$, i.e., $V = V_2$ or $V = V_1$ (since v^+ generates V). Similarly, it is clear that each proper submodule of V lies in the sum of weight spaces other than V_λ, so the sum of all such submodules is still proper.

Finally, (e) is clear. $\square$

Corollary. *Let V be as in the theorem. Suppose further that V is an irreducible L-module. Then v^+ is the unique maximal vector in V, up to nonzero scalar multiples.*

Proof. If w^+ is another maximal vector, then $\mathfrak{U}(L).w^+ = V$ (since V is irreducible). Therefore the theorem applies equally to v^+ and to w^+. If w^+ has weight λ', then $\lambda' \prec \lambda$ and $\lambda \prec \lambda'$ (by part (b)), requiring $\lambda = \lambda'$. But then (by part (c)) w^+ is proportional to v^+. $\square$

20.3. Existence and uniqueness theorems

We want to show that for each $\lambda \in H^*$, there exists one and (up to isomorphism) only one irreducible standard cyclic L-module of highest weight λ, which may be infinite dimensional. The *uniqueness* part is not difficult (the argument is similar to that used in proving Theorem 14.2, but less complicated).

Theorem A. *Let V, W be standard cyclic modules of highest weight λ. If V and W are irreducible, then they are isomorphic.*

Proof. Form the L-module $X = V \oplus W$. If v^+, w^+ are respective maximal vectors of weight λ in V, W, let $x^+ = (v^+, w^+) \in X$, so x^+ is a maximal vector of weight λ. Let Y be the L-submodule of X generated by x^+ (Y is standard cyclic), and let $p: Y \to V$, $p': Y \to W$ be the maps induced by projecting X onto its first and second factors. It is obvious that p, p' are L-module homomorphisms; since $p(x^+) = v^+$, $p'(x^+) = w^+$, it is also clear that Im $p = V$, Im $p' = W$. We claim that *each of p, p' is injective* (so that Y is isomorphic as L-module to each of V, W, thereby proving the theorem). Indeed, the kernel of p' is $V \cap Y$ (V being identified with the set of pairs $(u, 0)$ in X), an L-submodule of V. Since V is irreducible, $V \cap Y = 0$ or else $V \cap Y = V$. The latter is impossible: $(v^+, 0) \in Y$ would imply that $(v^+, 0)$ is proportional to (v^+, w^+) (Theorem 20.2(c)), which is absurd. Therefore p' (and similarly p) is injective. $\square$

Next we consider the *existence* question. Leaving aside all mention of irreducibility, the question remains: How can we construct any standard cyclic modules at all? There are two illuminating ways to proceed, which lead to the same results.

First we look at an *induced module* construction (which is similar to a technique used in the representation theory of finite groups). This is suggested by the observation that a standard cyclic module, viewed as B-module ($B = B(\Delta)$, as above), contains a one dimensional submodule spanned by the given maximal vector. Accordingly, we begin with a one dimensional vector space D_λ, having v^+ as basis, and define an action of B on D_λ by the rule $(h + \sum_{\alpha \succ 0} x_\alpha).v^+ = h.v^+ = \lambda(h)v^+$, for fixed $\lambda \in H^*$. The reader can quickly convince himself that this makes D_λ a B-module. Of course, D_λ is equally well a $\mathfrak{U}(B)$-module, so it makes sense to form the tensor product $Z(\lambda) = \mathfrak{U}(L) \otimes_{\mathfrak{U}(B)} D_\lambda$, which becomes a $\mathfrak{U}(L)$-module under the natural (left) action of $\mathfrak{U}(L)$.

We claim that $Z(\lambda)$ *is standard cyclic of weight* λ. On the one hand, $1 \otimes v^+$ evidently generates $Z(\lambda)$. On the other hand, $1 \otimes v^+$ is nonzero, because $\mathfrak{U}(L)$ is a free $\mathfrak{U}(B)$-module (Corollary D of Theorem 17.3) with basis consisting of 1 along with the various monomials $y_{\beta_1}^{i_1} \ldots y_{\beta_m}^{i_m}$. Therefore $1 \otimes v^+$ is a maximal vector of weight λ. Call it v^+ for brevity.

This construction also makes it clear that, if $N^- = \coprod_{\alpha \prec 0} L_\alpha$, then $Z(\lambda)$ viewed as $\mathfrak{U}(N^-)$-module is isomorphic to $\mathfrak{U}(N^-)$ itself. To be precise, $\mathfrak{U}(L) \cong \mathfrak{U}(N^-) \otimes \mathfrak{U}(B)$ (PBW Theorem), so that $Z(\lambda) \cong \mathfrak{U}(N^-) \otimes \mathsf{F}$ (as left $\mathfrak{U}(N^-)$-modules).

It is also possible to construct $Z(\lambda)$ by "generators and relations". For this choose, as before, nonzero elements $x_\alpha \in L_\alpha$ ($\alpha \succ 0$), and let $I(\lambda)$ be the left ideal in $\mathfrak{U}(L)$ generated by all x_α ($\alpha \succ 0$) along with all $h_\alpha - \lambda(h_\alpha).1$ ($\alpha \in \Phi$). Notice that these generators of $I(\lambda)$ annihilate the maximal vector v^+ of $Z(\lambda)$, so $I(\lambda)$ also does, and there is a canonical homomorphism of left $\mathfrak{U}(L)$-modules $\mathfrak{U}(L)/I(\lambda) \to Z(\lambda)$ sending the coset of 1 onto the maximal vector v^+. Again using our PBW basis of $\mathfrak{U}(L)$, we see that this map sends the cosets of $\mathfrak{U}(B)$ onto the line $\mathsf{F}v^+$, so it follows that the map is one-one. In other words, $Z(\lambda)$ is isomorphic to $\mathfrak{U}(L)/I(\lambda)$.

Theorem B. *Let* $\lambda \in H^*$. *Then there exists an irreducible standard cyclic module* $V(\lambda)$ *of weight* λ.

Proof. $Z(\lambda)$ (constructed above) is standard cyclic of weight λ, and has a unique maximal submodule $Y(\lambda)$ (Theorem 20.2(d)). Therefore, $V(\lambda) = Z(\lambda)/Y(\lambda)$ is irreducible and standard cyclic of weight λ (Theorem 20.2(e)). ☐

Two basic problems remain: (1) Decide which of the $V(\lambda)$ are finite dimensional. (2) Determine for such $V(\lambda)$ exactly which weights μ occur and with what multiplicity. The following sections are devoted to solving these problems.

Exercises

1. If V is an arbitrary L-module, then the sum of its weight spaces is direct.
2. (a) If V is an irreducible L-module having at least one (nonzero) weight space, prove that V is the direct sum of its weight spaces.

(b) Let V be an irreducible L-module. Then V has a (nonzero) weight space if and only if $\mathfrak{U}(H).v$ is finite dimensional for all $v \in V$, or if and only if $\mathfrak{A}.v$ is finite dimensional for all $v \in V$ (where $\mathfrak{A} = $ subalgebra with 1 generated by an arbitrary $h \in H$ in $\mathfrak{U}(H)$).

(c) Let $L = \mathfrak{sl}(2, \mathsf{F})$, with standard basis (x, y, h). Show that $1 - x$ is not invertible in $\mathfrak{U}(L)$, hence lies in a maximal left ideal I of $\mathfrak{U}(L)$. Set $V = \mathfrak{U}(L)/I$, so V is an irreducible L-module. Prove that the images of 1, h, $h^2, \ldots$ are all linearly independent in V (so dim $V = \infty$), using the fact that

$$(x-1)^r h^s \equiv \begin{cases} 0 \;(\mathrm{mod}\; I), & r > s \\ (-2)^r r! \cdot 1 \;(\mathrm{mod}\; I), & r = s. \end{cases}$$

Conclude that V has no (nonzero) weight space.

3. Describe weights and maximal vectors for the natural representations of the linear Lie algebras of types $A_\ell - D_\ell$ described in (1.2).

4. Let $L = \mathfrak{sl}(2, \mathsf{F})$, $\lambda \in H^*$. Prove that the module $Z(\lambda)$ for $\lambda = \lambda(h)$ constructed in Exercise 7.7 is isomorphic to the module $Z(\lambda)$ constructed in (20.3). Deduce that dim $V(\lambda) < \infty$ if and only if $\lambda(h)$ is a nonnegative integer.

5. If $\mu \in H^*$, define $\mathscr{P}(\mu)$ to be the number of distinct sets of nonnegative integers k_α ($\alpha \succ 0$) for which $\mu = \sum_{\alpha \succ 0} k_\alpha \alpha$. Prove that dim $Z(\lambda)_\mu = \mathscr{P}(\lambda - \mu)$, by describing a basis for $Z(\lambda)_\mu$.

6. Prove that the left ideal $I(\lambda)$ introduced in (20.3) is already generated by the elements x_α, $h_\alpha - \lambda(h_\alpha).1$ for α simple.

7. Prove, without using the induced module construction in (20.3), that $I(\lambda) \cap \mathfrak{U}(N^-) = 0$, in particular that $I(\lambda)$ is properly contained in $\mathfrak{U}(L)$. [Show that the analogous left ideal $I'(\lambda)$ in $\mathfrak{U}(B)$ is proper, while $I(\lambda) = \mathfrak{U}(N^-)I'(\lambda)$ by PBW.]

8. For each positive integer d, prove that the number of distinct irreducible L-modules $V(\lambda)$ of dimension $\le d$ is finite. Conclude that the number of nonisomorphic L-modules of dimension $\le d$ is finite. [If dim $V(\lambda) < \infty$, view $V(\lambda)$ as S_α-module for each $\alpha \succ 0$; notice that $\lambda(h_\alpha) \in \mathbf{Z}$, and that $V(\lambda)$ includes an S_α-submodule of dimension $\lambda(h_\alpha) + 1$.]

9. Verify the following description of the unique maximal submodule $Y(\lambda)$ of $Z(\lambda)$ (20.3): If $v \in Z(\lambda)_\mu$, $\lambda - \mu = \sum_{\alpha \succ 0} c_\alpha \alpha$ ($c_\alpha \in \mathbf{Z}^+$), observe that $\prod_{\alpha \succ 0} x_\alpha^{c_\alpha}.v$ has weight λ (the positive roots in any fixed order), hence is a scalar multiple of the maximal vector v^+. If this multiple is 0 for every possible choice of the c_α (cf. Exercise 5), prove that $v \in Y(\lambda)$. Conversely, prove that $Y(\lambda)$ is the span of all such weight vectors v for weights $\mu \ne \lambda$.

10. A maximal vector w^+ of weight μ in $Z(\lambda)$ induces an L-module homomorphism $\phi: Z(\mu) \to Z(\lambda)$, with Im ϕ the submodule generated by w^+. Prove that ϕ is injective.

11. Let V be an arbitrary finite dimensional L-module, $\lambda \in H^*$. Construct in the L-module $W = Z(\lambda) \otimes V$ a chain of submodules $W = W_1 \supset W_2$

$\supset \ldots \supset W_{n+1} = 0$ $(n = \dim V)$ so that W_i/W_{i+1} is isomorphic to $Z(\lambda + \lambda_i)$, where the weights of V in suitable order (multiplicities counted) are $\lambda_1, \ldots, \lambda_n$.

Notes

Lemire [1] treats the question of existence of weight spaces in infinite dimensional modules; Exercise 2 is due to him. The modules $Z(\lambda)$ are explored in detail by Verma [1], and more recently by Bernstein, Gel'fand, Gel'fand [1], [2]. Verma shows in particular that the space of L-homomorphisms $Z(\mu) \to Z(\lambda)$ is either zero or 1-dimensional over F, and obtains a sufficient condition for the second of these, while the latter authors prove the (conjectured) necessity of the condition.

21. Finite dimensional modules

21.1. Necessary condition for finite dimension

Suppose V is a finite dimensional irreducible L-module. Then V has at least one maximal vector, of uniquely determined weight λ, and the submodule it generates must be all of V (by irreducibility). Therefore, V is isomorphic to $V(\lambda)$ (Theorems A and B of (20.3)).

For each simple root α_i, let S_i $(= S_{\alpha_i})$ be the corresponding copy of $\mathfrak{sl}(2, \mathsf{F})$ in L. Then $V(\lambda)$ is also a (finite dimensional) module for S_i, and a maximal vector for L is also a maximal vector for S_i. In particular, if there is a maximal vector of weight λ, then the weight for the CSA $H_i \subset S_i$ is completely determined by the scalar $\lambda(h_i)$, $h_i = h_{\alpha_i}$. But this forces $\lambda(h_i)$ to be a nonnegative integer, thanks to Theorem 7.2. This proves:

Theorem. *If V is a finite dimensional irreducible L-module of highest weight λ, then $\lambda(h_i)$ is a nonnegative integer* $(1 \leq i \leq \ell)$. $\square$

More generally, it follows from (7.2) that if V is any finite dimensional L-module, μ a weight of V, then $\mu(h_i) = \langle \mu, \alpha_i \rangle \in \mathbf{Z}$, for $1 \leq i \leq \ell$. Accordingly, the weights occurring in a finite dimensional module are also "weights" in the sense of the abstract theory developed in §13, so all results proved there are available from now on. Notice that, in the language of §13, the highest weight λ of $V(\lambda)$ (in case $\dim V(\lambda) < \infty$) is *dominant*. To avoid ambiguity, we shall continue to allow any element of H^* to be called a weight, whereas a linear function λ for which all $\lambda(h_i)$ (hence all $\lambda(h_\alpha)$) are integral will be called **integral**. If all $\lambda(h_i)$ are nonnegative integers, then we call λ **dominant integral**. The set Λ of all integral linear functions is therefore a lattice in H^* (or equally well, in the real euclidean space generated by the roots), which includes the root lattice. As in §13, the set of dominant integral linear functions is denoted Λ^+.

One further bit of notation will be handy: *If V is an L-module, let $\Pi(V)$ denote the set of all its weights. For $V = V(\lambda)$, write instead $\Pi(\lambda)$.*

21.2. Sufficient condition for finite dimension

Theorem. *If $\lambda \in H^*$ is dominant integral, then the irreducible L-module $V = V(\lambda)$ is finite dimensional, and its set of weights $\Pi(\lambda)$ is permuted by $\mathcal{W}$, with $\dim V_\mu = \dim V_{\sigma\mu}$ for $\sigma \in \mathcal{W}$.*

Corollary. *The map $\lambda \mapsto V(\lambda)$ induces a one-one correspondence between Λ^+ and the isomorphism classes of finite dimensional irreducible L-modules.*

Proof of Corollary. This follows from the theorem, in view of Theorem 21.1 and Theorems A, B of (20.3). $\square$

Proof of Theorem. It will be convenient to write down first some information about commutators in $\mathfrak{U}(L)$. Fix standard generators $\{x_i, y_i\}$ of L.

Lemma. *The following identities hold in $\mathfrak{U}(L)$, for $k \geq 0$, $1 \leq i, j \leq \ell$:*

(a) $[x_j, y_i^{k+1}] = 0$ *when $i \neq j$;*
(b) $[h_j, y_i^{k+1}] = -(k+1)\alpha_i(h_j)y_i^{k+1}$;
(c) $[x_i, y_i^{k+1}] = -(k+1)y_i^k(k \cdot 1 - h_i)$.

Proof. (a) follows from the fact (Lemma 10.1) that $\alpha_j - \alpha_i$ is not a root when $i \neq j$.

For (b), use induction on k, the case $k = 0$ being $[h_j, y_i] = -\alpha_i(h_j)y_i$ (cf. (18.1)). In general, the left side equals $h_j y_i^{k+1} - y_i^{k+1} h_j = (h_j y_i^k - y_i^k h_j)y_i + y_i^k(h_j y_i - y_i h_j) = -k\alpha_i(h_j)y_i^k y_i + y_i^k(-\alpha_i(h_j)y_i) = -(k+1)\alpha_i(h_j)y_i^{k+1}$, using the induction hypothesis in the next-to-last step.

For (c) write $[x_i, y_i^{k+1}] = x_i y_i^{k+1} - y_i^{k+1} x_i = [x_i, y_i]y_i^k + y_i[x_i, y_i^k] = h_i y_i^k + y_i[x_i, y_i^k]$, and use induction on k along with formula (b) (for k in place of $k+1$). $\square$

The proof of the theorem will be carried out in steps. The idea is to show that the set of weights of V is permuted by $\mathcal{W}$, hence is finite (cf. the proof of Theorem 18.3). It is convenient to denote the representation of L afforded by V by $\phi: L \to \mathfrak{gl}(V)$. Fix a maximal vector v^+ of V (of weight λ), and set $m_i = \lambda(h_i)$, $1 \leq i \leq \ell$. The m_i are nonnegative integers, by assumption.

(1) $y_i^{m_i+1}.v^+ = 0$. Let $w = y_i^{m_i+1}.v^+$. By part (a) of the lemma, when $i \neq j$, $x_j.w = 0$. On the other hand, parts (b) and (c) of the lemma show that $x_i y_i^{m_i+1}.v^+ = y_i^{m_i+1} x_i.v^+ - (m_i+1)y_i^{m_i}.(m_i v^+ - m_i v^+) = 0$, so $x_i.w = 0$. If w were nonzero, it would therefore be a maximal vector in V of weight $\lambda - (m_i+1)\alpha_i \neq \lambda$, contrary to Corollary 20.2.

(2) *For $1 \leq i \leq \ell$, V contains a nonzero finite dimensional S_i-module.* The subspace spanned by v^+, $y_i.v^+$, $y_i^2.v^+$, ..., $y_i^{m_i}.v^+$ is stable under y_i, according to step (1). It is also stable under h_i, since each of these vectors belongs to a weight space of V; so it is stable under x_i, by part (c) of the lemma (and induction on the superscript k).

(3) *V is the sum of finite dimensional S_i-submodules.* If V' denotes the

sum of all such submodules of V, then V' is nonzero by step (2). On the other hand, let W be any finite dimensional S_i-submodule of V. The span of all subspaces $x_\alpha W$ ($\alpha \in \Phi$) is evidently finite dimensional, as well as S_i-stable. Therefore V' is stable under L, and $V' = V$ because V is irreducible.

(4) *For $1 \leq i \leq \ell$, $\phi(x_i)$ and $\phi(y_i)$ are locally nilpotent endomorphisms of V* (cf. (18.3)). Indeed, if $v \in V$, then v lies in a finite sum of finite dimensional S_i-submodules (hence in a finite dimensional S_i-submodule), by (3). On such a module $\phi(x_i)$ and $\phi(y_i)$ are nilpotent (cf. (6.4)).

(5) $s_i = exp\ \phi(x_i)\ exp\ \phi(-y_i)\ exp\ \phi(x_i)$ *is a well-defined automorphism of V.* This follows at once from (4) (cf. (18.3) again!).

(6) *If μ is any weight of V, then $s_i(V_\mu) = V_{\sigma_i \mu}$ (σ_i = reflection relative to α_i).* V_μ lies in a finite dimensional S_i-submodule V' (cf. step (3)), and $s_i|_{V'}$ is the same as the automorphism τ constructed in (7.2); the claim now follows from the discussion in (7.2).

(7) *The set of weights $\Pi(\lambda)$ is stable under $\mathscr{W}$, and $\dim V_\mu = \dim V_{\sigma\mu}$* ($\mu \in \Pi(\lambda)$, $\sigma \in \mathscr{W}$). Since $\mathscr{W}$ is generated by $\sigma_1, \ldots, \sigma_\ell$ (Theorem 10.3(d)), this follows from (6).

(8) $\Pi(\lambda)$ *is finite.* It is clear from Lemma 13.2B that the set of $\mathscr{W}$-conjugates of all dominant integral linear functions $\mu \prec \lambda$ is finite. But $\Pi(\lambda)$ is included in this set, thanks to Theorem 20.2, combined with step (7).

(9) *$\dim V$ is finite.* We know from Theorem 20.2(c) that $\dim V_\mu$ is finite for all $\mu \in \Pi(\lambda)$. Combined with step (8), this proves our assertion.   ☐

21.3. Weight strings and weight diagrams

We remain in the finite dimensional situation, $V = V(\lambda)$, $\lambda \in \Lambda^+$. Let $\mu \in \Pi(\lambda)$, $\alpha \in \Phi$. Lemma 20.1 shows that the subspace W of V spanned by all weight spaces $V_{\mu + i\alpha}$ ($i \in \mathbf{Z}$) is invariant under S_α. In view of (7.2) and Weyl's Theorem on complete reducibility, the weights in $\Pi(\lambda)$ of the form $\mu + i\alpha$ must form a connected string (the **α-string through μ**, generalizing the notion of α-string through β for roots in the adjoint representation). Moreover, the reflection σ_α reverses this string. If the string consists of $\mu - r\alpha, \ldots, \mu, \ldots, \mu + q\alpha$, it follows that $r - q = \langle \mu, \alpha \rangle$. This proves the following result, in view of (13.4).

Proposition. *If $\lambda \in \Lambda^+$, the set $\Pi(\lambda)$ is saturated in the sense of (13.4). In particular, the necessary and sufficient condition for $\mu \in \Lambda$ to belong to $\Pi(\lambda)$ is that μ and all its $\mathscr{W}$-conjugates be $\prec \lambda$.*   ☐

All of this can be visualized quite easily when rank $\Phi \leq 2$, if we draw a weight diagram. For example, let $L = \mathfrak{sl}(3, \mathsf{F})$ (type A_2), with fundamental dominant integral linear functions (13.2) λ_1, λ_2. The weight diagram for $V(\lambda)$, $\lambda = 4\lambda_1 + 3\lambda_2$, is given in Figure 1. The dots indicate which weights occur. The multiplicities in this case are also indicated, increasing from one to four (passing from outer "shell" to inner "shell", the multiplicity $\dim V_\mu$ increases steadily by one until the shells become triangles, at which point multiplicity stabilizes). The simple behavior of these multiplicities is a special

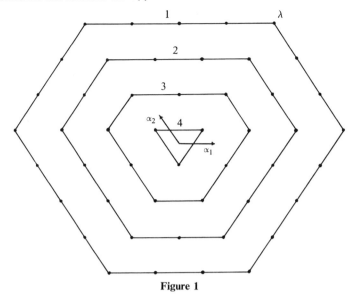

Figure 1

fact about type A_2 (Antoine, Speiser [1]). For other root systems, the situation can be much more complicated; see §22 below for a more detailed discussion of multiplicities.

21.4. Generators and relations for V(λ)

It is possible to describe more precisely the passage from $Z(\lambda)$ to its homomorphic image $V(\lambda)$, when λ is dominant integral. This is not needed elsewhere in the text, but is of independent interest. In effect we shall rework some of the proof of Theorem 21.2.

Recall from (20.3) that $Z(\lambda)$ is isomorphic to $\mathfrak{U}(L)/I(\lambda)$, where $I(\lambda)$ is the left ideal of $\mathfrak{U}(L)$ generated by all x_α ($\alpha \succ 0$), and by all $h_\alpha - \lambda(h_\alpha) \cdot 1$ ($\alpha \in \Phi$). Equivalently, $I(\lambda)$ is the annihilator of a maximal vector in $Z(\lambda)$. Now fix a dominant integral linear function λ, and let $J(\lambda)$ be the left ideal in $\mathfrak{U}(L)$ which annihilates a maximal vector of $V(\lambda)$. The inclusion $I(\lambda) \subset J(\lambda)$ induces the canonical map $Z(\lambda) = \mathfrak{U}(L)/I(\lambda) \to V(\lambda) \cong \mathfrak{U}(L)/J(\lambda)$. From the proof of Theorem 21.2 we recall that $y_i^{m_i+1} \in J(\lambda)$, $1 \le i \le \ell$, where $m_i = \langle \lambda, \alpha_i \rangle$.

Theorem. *Let $\lambda \in \Lambda^+$, $m_i = \langle \lambda, \alpha_i \rangle$ ($1 \le i \le \ell$). Then $J(\lambda)$ is generated by $I(\lambda)$ along with all $y_i^{m_i+1}$ ($1 \le i \le \ell$).*

Proof. Suppose we can show that $V'(\lambda) = \mathfrak{U}(L)/J'(\lambda)$ is *finite dimensional*, where $J'(\lambda)$ is the left ideal generated by $I(\lambda)$ along with all $y_i^{m_i+1}$. Since $V'(\lambda)$ is standard cyclic (or 0), it must be irreducible (or 0) (Theorem 20.2(d), with Weyl's Theorem on complete reducibility). But $J'(\lambda) \subset J(\lambda)$ implies that $V(\lambda)$ is a homomorphic image of $V'(\lambda)$, forcing $V'(\lambda) \cong V(\lambda)$, whence $J'(\lambda) = J(\lambda)$ as desired.

To show that $V'(\lambda)$ is finite dimensional, it would in turn suffice to show that it is a sum of finite dimensional S_i-submodules ($1 \le i \le \ell$), because then the proof of Theorem 21.2 would go through exactly as before. For this *it is enough to show that each y_i is locally nilpotent on $V'(\lambda)$* (this is of course obvious already for the x_i, since we cannot have $\mu + k\alpha_i \prec \lambda$ for all $k \ge 0$). By hypothesis, the coset of 1 in $V'(\lambda)$ is killed by a suitable power of y_i (namely, $m_i + 1$). We know (Theorem 20.2) that $V'(\lambda)$ is spanned by the cosets of all $y_{i_1} \ldots y_{i_t}$ ($1 \le i_j \le \ell$). The following lemma implies that if the coset of this monomial is killed (i.e. sent into $J'(\lambda)$) by y_i^k, then the coset of the longer monomial $y_{i_0} y_{i_1} \ldots y_{i_t}$ is killed by y_i^{k+3}. Induction on length of monomials, starting at 1, then proves the local nilpotence of y_i.

Lemma. *Let $\mathfrak{A}$ be an associative algebra over F, $y, z \in \mathfrak{A}$. Then $[y^k, z] =$*
$$\binom{k}{1} [y, z] y^{k-1} + \binom{k}{2} [y, [y, z]] y^{k-2} + \ldots + [y, [y, \ldots, [y, z] \ldots] \ldots].$$

Proof. Use induction on k, the case $k = 1$ being the identity $[y, z] = [y, z]$. The induction step is easy, and is left to the reader. $\square$

To apply the lemma, take $\mathfrak{A}$ to be $\mathfrak{U}(L)$, and take y, z to be root vectors belonging to two negative roots. We know that $(\text{ad } y)^4(z) = 0$, since root strings have length at most 4, so the identity obtained in the lemma reduces to: $[y^k, z] = k[y, z] y^{k-1} + \binom{k}{2} [y, [y, z]] y^{k-2} + \binom{k}{3} [y, [y, [y, z]]] y^{k-3}$. $\square$

Exercises

1. The reader can check that we have not yet used the *simple transitivity* of $\mathscr{W}$ on bases of Φ (Theorem 10.3(e)), only the transitivity. Use representation theory to obtain a new proof, as follows: There exists a finite dimensional irreducible module $V(\lambda)$ for which all $\langle \lambda, \alpha \rangle$ ($\alpha \in \Delta$) are distinct and positive. If $\sigma \in \mathscr{W}$ permutes Δ, then $\sigma\lambda = \lambda$, forcing $\sigma = 1$.
2. Draw the weight diagram for the case B_2, $\lambda = \lambda_1 + \lambda_2$ (notation of Chapter III).
3. Let $\lambda \in \Lambda^+$. Prove that 0 occurs as a weight of $V(\lambda)$ if and only if λ is a sum of roots.
4. Recall the module $Z(\lambda)$ constructed in (20.3). Use Lemma 21.2 to find certain maximal vectors in $Z(\lambda)$, when $\lambda \in \Lambda$: the coset of $y_i^{m_i+1}$, $m_i = \langle \lambda, \alpha_i \rangle$, is a maximal vector provided m_i is nonnegative (Cf. Exercise 7.7.)
5. Let V be a *faithful* finite dimensional L-module, $\Lambda(V)$ the subgroup of Λ generated by the weights of V. Then $\Lambda(V) \supset \Lambda_r$. Show that every subgroup of Λ including Λ_r is of this type.
6. If $V = V(\lambda)$, $\lambda \in \Lambda^+$, prove that V^* is isomorphic (as L-module) to $V(-\sigma\lambda)$, where $\sigma \in \mathscr{W}$ is the unique element of $\mathscr{W}$ sending Δ to $-\Delta$ (Exercise 10.9, cf. Exercise 13.4).

7. Let $V = V(\lambda)$, $W = V(\mu)$, with λ, $\mu \in \Lambda^+$. Prove that $\Pi(V \otimes W) = \{\nu + \nu' | \nu \in \Pi(\lambda), \nu' \in \Pi(\mu)\}$ and that dim $(V \otimes W)_{\nu+\nu'}$ equals

$$\sum_{\pi + \pi' = \nu + \nu'} \dim V_\pi \cdot \dim W_{\pi'}.$$

In particular, $\lambda + \mu$ occurs with multiplicity one, so $V(\lambda+\mu)$ occurs exactly once as a direct summand of $V \otimes W$.

8. Let $\lambda_1, \ldots, \lambda_\ell$ be the fundamental dominant weights for the root system Φ of L (13.1). Show how to construct an arbitrary $V(\lambda)$, $\lambda \in \Lambda^+$, as a direct summand in a suitable tensor product of modules $V(\lambda_1), \ldots, V(\lambda_\ell)$ (repetitions allowed).

9. Prove Lemma 21.4 and deduce Lemma 21.2 from it.

10. Let $L = \mathfrak{sl}(\ell+1, \mathsf{F})$, with CSA $H = \mathfrak{d}(\ell+1, \mathsf{F}) \cap L$. Let $\mu_1, \ldots, \mu_{\ell+1}$ be the coordinate functions on H, relative to the standard basis of $\mathfrak{gl}(\ell+1, \mathsf{F})$. Then $\sum \mu_i = 0$, and $\mu_1, \ldots, \mu_\ell$ form a basis of H^*, while the set of $\alpha_i = \mu_i - \mu_{i+1}$ ($1 \le i \le \ell$) is a base Δ for the root system Φ. Verify that $\mathscr{W}$ acts on H^* by permuting the μ_i; in particular, the reflection with respect to α_i interchanges μ_i, μ_{i+1} and leaves the other μ_j fixed. Then show that the fundamental dominant weights relative to Δ are given by $\lambda_k = \mu_1 + \ldots + \mu_k$ ($1 \le k \le \ell$).

11. Let $V = \mathsf{F}^{\ell+1}$, $L = \mathfrak{sl}(V)$. Fix the CSA H and the base $\Delta = (\alpha_1, \ldots, \alpha_\ell)$ of Φ as in Exercise 10. The purpose of this exercise is to construct irreducible L-modules V_k ($1 \le k \le \ell$) of highest weight λ_k.
   (a) For $k = 1$, $V_1 = V$ is irreducible of highest weight λ_1.
   (b) In the k-fold tensor product $V \otimes \ldots \otimes V$, $k \ge 2$, define V_k to be the subspace of **skew-symmetric tensors**: If $(v_1, \ldots, v_{\ell+1})$ is the canonical basis of V, V_k has basis consisting of the $\binom{\ell+1}{k}$ vectors

(*)
$$[v_{i_1}, \ldots, v_{i_k}] = \sum_{\pi \in \mathscr{S}_k} sn(\pi) v_{\pi(i_1)} \otimes \ldots \otimes v_{\pi(i_k)},$$

where $i_1 < i_2 < \ldots < i_k$. Show that (*) is of weight $\mu_{i_1} + \ldots + \mu_{i_k}$.
   (c) Prove that L leaves the subspace V_k invariant and that all the weights $\mu_{i_1} + \ldots + \mu_{i_k}$ ($i_1 < \ldots < i_k$) are distinct and conjugate under $\mathscr{W}$. Conclude that V_k is irreducible, of highest weight λ_k. (Cf. Exercise 13.13.)

Notes

Theorem 21.4 is more or less well known; our treatment is based on an addendum to the thesis of Verma [1], cf. also Harish-Chandra [1]. Weight diagrams for A_2 appear in Antoine, Speiser [1]; see also Kolman, Belinfante [1], and Samelson [1].

22. Multiplicity formula

All modules considered in this section are finite dimensional.

If $\mu \in H^*$ is an integral linear function, define the **multiplicity** of μ in $V(\lambda)$, $\lambda \in \Lambda^+$, to be $m_\lambda(\mu) = \dim V(\lambda)_\mu$ ($= 0$ in case μ is not a weight of

$V(\lambda)$). When λ is fixed we write simply $m(\mu)$. Our aim is to derive Freuden-thal's recursion formula for $m_\lambda(\mu)$, by computing the trace of a Casimir-type element on $V(\lambda)_\mu$; using the fact that the element in question acts as a non-zero scalar on $V(\lambda)$, we can recover in this way dim $V(\lambda)_\mu$.

22.1. A universal Casimir element

Recall from (6.2) the notion of Casimir element c_ϕ of a representation of L, which was used to prove Weyl's Theorem on complete reducibility. Now that $\mathfrak{U}(L)$ is available we can make a "universal" construction of this sort.

Begin with the adjoint representation of L, whose trace form is just the Killing form κ. It is easy to deduce from §8 a natural construction of dual bases relative to κ. If α, β are arbitrary linear functions on H, we know that L_α is orthogonal to L_β except when $\beta = -\alpha$ (Proposition 8.1). We also know that the restriction of κ to H is nondegenerate. Therefore we can proceed as follows. Choose any basis of H, say the standard one $(h_1, \ldots, h_\ell)$ (relative to Δ), and let $(k_1, \ldots, k_\ell)$ be the dual basis of H, relative to the restriction of κ to H. Next choose nonzero x_α in each L_α ($\alpha \in \Phi$), and let z_α be the (unique) element of $L_{-\alpha}$ satisfying $\kappa(x_\alpha, z_\alpha) = 1$. By the remarks above, the bases $(h_i, 1 \le i \le \ell; x_\alpha, \alpha \in \Phi)$ and $(k_i, 1 \le i \le \ell; z_\alpha, \alpha \in \Phi)$ are dual relative to κ. *A word of caution, however*: The pair x_α, z_α must not be confused with our customary choice of x_α, y_α such that $[x_\alpha y_\alpha] = h_\alpha$. Rather, we have here $[x_\alpha z_\alpha] = t_\alpha = [(\alpha, \alpha)/2]h_\alpha$ (Proposition 8.3(c)).

By definition, a Casimir element for ad is the endomorphism of L given by $c_{\mathrm{ad}} = \sum_{i=1}^{\ell} \mathrm{ad}\, h_i \,\mathrm{ad}\, k_i + \sum_{\alpha \in \Phi} \mathrm{ad}\, x_\alpha \,\mathrm{ad}\, z_\alpha$. This construction might suggest to the reader consideration of the element $c_L = \sum_{i=1}^{\ell} h_i k_i + \sum_{\alpha \in \Phi} x_\alpha z_\alpha \in \mathfrak{U}(L)$. If ad is extended (uniquely) to a homomorphism of associative algebras ad: $\mathfrak{U}(L) \to \mathrm{End}\, L$, then ad c_L is none other than c_{ad}. For this reason we call c_L a **universal Casimir element** of L. The argument in (6.2) shows that for any representation ϕ of L, $\phi(c_L)$ commutes with $\phi(L)$, hence acts as a scalar if ϕ is irreducible.

To make sure that $\phi(c_L) \ne 0$ when $\phi \ne 0$, we investigate how $\phi(c_L)$ is related to a Casimir element c_ϕ. This is easy to see in case L is simple, so we treat this case first (cf. Exercise 6.6).

Lemma. *Let L be a simple Lie algebra. If $f(x, y)$ and $g(x, y)$ are non-degenerate symmetric, associative bilinear forms on L, then there is a nonzero scalar a such that $f(x, y) = ag(x, y)$ for all $x, y \in L$.*

Proof. Each form (being nondegenerate) sets up a natural vector space isomorphism of L onto L^*, via $x \mapsto s$, where $s(y) = f(x, y)$ or $g(x, y)$. The associativity guarantees that these are actually isomorphisms of L-modules (recall from (6.1) how L^* is made into an L-module). Combining one of these maps with the inverse of the other therefore sets up an L-module isomorphism $\pi: L \to L$. But L is an irreducible L-module (being simple), so

π is a scalar multiplication, thanks to Schur's Lemma. In other words, we have $0 \neq a \in F$ such that if $f(x, y) = g(z, y)$ (for all $y \in L$), then $z = ax$. □

Let $\phi\colon L \to \mathfrak{gl}(V)$ be a representation of L, L simple. If $\phi(L) = 0$, everything is clear. Otherwise ϕ is *faithful* (since Ker ϕ is an ideal of L), so the form $f(x, y) = Tr\ (\phi(x)\phi(y))$ on L is nondegenerate, as well as associative. The Killing form has the same properties, so it must be a nonzero multiple af (by the lemma). In particular, given one basis of L, the dual basis relative to κ is gotten by multiplying the dual basis vectors relative to f by $1/a$. This shows that $\phi(c_L) = (1/a)c_\phi$. In words, *the ordinary Casimir element of ϕ is a nonzero multiple of the image of the universal Casimir element.*

Finally, let L be semisimple. We observed in (5.2) that distinct simple ideals of L are orthogonal to each other, relative to κ. This makes it clear that the dual bases selected above can be chosen to be unions of analogous dual bases for the simple components of L (relative to their Killing forms, which are gotten by restricting κ). Therefore $c_L = c_{L_1} + \ldots + c_{L_t}$ ($L = L_1 \oplus \ldots \oplus L_t$), and if ϕ is a representation of L, each $\phi(c_{L_i})$ is proportional to the corresponding c_{ϕ_i} ($\phi_i = \phi|_{L_i}$), where ϕ_i is either trivial or faithful for each i. So $\phi(c_L)$ is again very closely related, though not necessarily proportional, to c_ϕ. In particular, this shows again that $\phi(c_L)$ commutes with $\phi(L)$; moreover, *it acts as a nonzero scalar when ϕ is irreducible* (unless $\phi(L)=0$). This last fact is all we really need below.

22.2. Traces on weight spaces

Fix an irreducible L-module $V = V(\lambda)$, $\lambda \in \Lambda^+$, and denote by ϕ the representation it affords. Fix also the dual bases of L relative to κ chosen in (22.1). In this subsection we are going to compute, for each weight μ of V, the trace on V_μ of the endomorphism $\phi(x_\alpha)\phi(z_\alpha)$. This makes sense, because $\phi(z_\alpha)$ maps V_μ into $V_{\mu - \alpha}$ and then $\phi(x_\alpha)$ maps $V_{\mu - \alpha}$ back into V_μ.

Since we are working with only one root α, we can utilize the representation theory of S_α (§7). Some modifications are needed, however, because our basis $(x_\alpha, z_\alpha, t_\alpha)$ is nonstandard; it is related to the standard basis $(x_\alpha, y_\alpha, h_\alpha)$ by $z_\alpha = [(\alpha, \alpha)/2]y_\alpha$, $t_\alpha = [(\alpha, \alpha)/2]h_\alpha$. Let $(v_0, v_1, \ldots, v_m)$ be the basis used in formulas (a)–(c), Lemma 7.2, for the irreducible S_α-module of highest weight m. It will be convenient to replace this basis by $(w_0, \ldots, w_m)$, where $w_i = i![(\alpha, \alpha)^i/2^i]v_i$. After making this substitution, we obtain:

(a') $t_\alpha.w_i = (m - 2i)\ [(\alpha, \alpha)/2]w_i$;

(b') $z_\alpha.w_i = w_{i+1}$, $(w_{m+1} = 0)$;

(c') $x_\alpha.w_i = i(m - i + 1)\ [(\alpha, \alpha)/2]w_{i-1}$, $(w_{-1} = 0)$.

Therefore:

(1) $x_\alpha z_\alpha.w_i = (m - i)\ (i + 1)\ [(\alpha, \alpha)/2]w_i$.

Now let μ be any weight of V for which $\mu + \alpha$ is *not* a weight. Then (21.3) the α-string of weights through μ consists of μ, $\mu - \alpha, \ldots, \mu - m\alpha$, where

$m = \langle \mu, \alpha \rangle$. *We keep μ, α, m fixed throughout the following discussion.* The representation of S_α on the sum of weight spaces $W = V_\mu + V_{\mu - \alpha} + \ldots + V_{\mu - m\alpha}$ is a direct sum of irreducible representations (Weyl's Theorem), each involving a string of weights stable under σ_α. To be more precise, let n_i ($0 \le i \le [m/2]$) denote the number of such constituents having highest weight $(\mu - i\alpha)(h_\alpha)$. Then $m(\mu - i\alpha) = n_0 + \ldots + n_i$, and in turn, $n_i = m(\mu - i\alpha) - m(\mu - (i-1)\alpha)$. This is shown schematically in Figure 1, for the case m even.

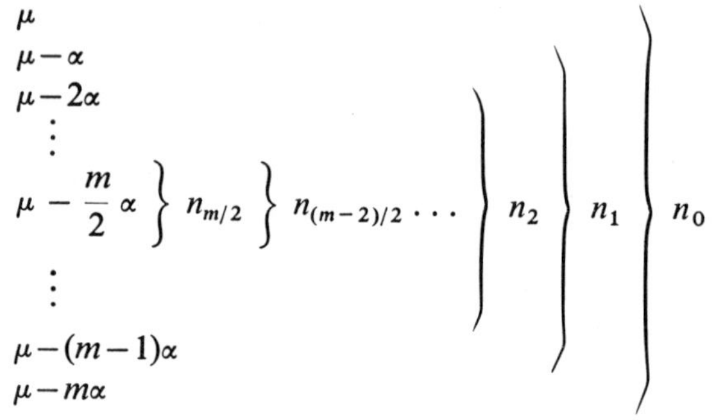

$$
\begin{matrix}
\mu \\
\mu - \alpha \\
\mu - 2\alpha \\
\vdots \\
\mu - \dfrac{m}{2}\alpha \quad \Big\} \, n_{m/2} \quad \Big\} \, n_{(m-2)/2} \, \cdots \quad \Big\} \, n_2 \quad \Big\} \, n_1 \quad \Big\} \, n_0 \\
\vdots \\
\mu - (m-1)\alpha \\
\mu - m\alpha
\end{matrix}
$$

Figure 1. (*m* even)

For each fixed k, $0 \le k \le m/2$, we want to calculate the trace of $\phi(x_\alpha)\phi(z_\alpha)$ on $V_{\mu - k\alpha}$. Let $0 \le i \le k$. In a typical irreducible S_α-summand of W having highest weight $m - 2i = (\mu - i\alpha)(h_\alpha)$, the weight space corresponding to $\mu - k\alpha$ is spanned by the vector w_{k-i} (in the above notation). Replacing m by $m - 2i$ and i by $k - i$ in formula (1), we obtain:

$$(2) \qquad \phi(x_\alpha)\phi(z_\alpha)w_{k-i} = (m - i - k)(k - i + 1)[(\alpha, \alpha)/2]w_{k-i}.$$

There are n_i S_α-summands of W having highest weight $m - 2i$, so the matrix of $\phi(x_\alpha)\phi(z_\alpha)$ (restricted to $V_{\mu - k\alpha}$) has n_i diagonal entries of the form given by (2), relative to a suitable basis of eigenvectors. Letting i range from 0 to k, we obtain for $\phi(x_\alpha)\phi(z_\alpha)$ a diagonal matrix of order $m(\mu - k\alpha) = n_0 + \ldots + n_k$, with trace:

$$(3) \qquad \sum_{i=0}^{k} n_i(m - i - k)(k - i + 1)(\alpha, \alpha)/2$$

$$= \sum_{i=0}^{k} (m(\mu - i\alpha) - m(\mu - (i-1)\alpha))(m - i - k)(k - i + 1)(\alpha, \alpha)/2$$

$$= \sum_{i=0}^{k} m(\mu - i\alpha)(m - 2i)(\alpha, \alpha)/2.$$

The last equality follows, because the coefficient of $m(\mu - i\alpha)$ is $(\alpha, \alpha)/2$ times $(m - i - k)(k - i + 1) - (m - i - k - 1)(k - i) = m - 2i$. (The reader should check directly the extreme case $i = k$.) Now recall that $m/2 = (\mu, \alpha)/(\alpha, \alpha)$. So (3) becomes:

$$(4) \qquad Tr_{V_{\mu - k\alpha}}\phi(x_\alpha)\phi(z_\alpha) = \sum_{i=0}^{k} m(\mu - i\alpha)(\mu - i\alpha, \alpha).$$

This takes care of the weights $\mu - k\alpha$ in the top half of the "ladder" (Figure 1). Since the reflection σ_α interchanges top and bottom, we can expect similar behavior; in particular, $m(\mu - i\alpha) = m(\mu - (m-i)\alpha)$ for $m/2 < i \leq m$. It is a straightforward matter to imitate the above reasoning for fixed k, $m/2 < k \leq m$, thereby obtaining:

$$(5) \qquad Tr_{V_{\mu - k\alpha}}\phi(x_\alpha)\phi(z_\alpha) = \sum_{i=0}^{m-k-1} m(\mu - i\alpha)\,(\mu - i\alpha, \alpha).$$

(We should sum to $m - k$, but that term is 0 already. Why?)

But notice that, for $m/2 < i \leq m$, $(\mu - i\alpha, \alpha) + (\mu - (m-i)\alpha, \alpha) = (2\mu - m\alpha, \alpha) = 0$, because $m = 2(\mu, \alpha)/(\alpha, \alpha)$. Therefore:

$$(6) \qquad m(\mu - i\alpha)\,(\mu - i\alpha, \alpha) + m(\mu - (m-i)\alpha)\,(\mu - (m-i)\alpha, \alpha) = 0.$$

This shows that certain pairs of summands may be added to (5): $k+1$ and $m - (k+1)$, $k+2$ and $m - (k+2)$, etc. (Note that for $m = 2i$ even, (6) forces $(\mu - i\alpha, \alpha) = 0$.) In other words, (5) *reduces to* (4), *for arbitrary k*.

Finally, if we wish to consider an arbitrary weight ν of V, we form the α-string through ν and let the final term $\nu + k\alpha$ play the role of μ in the above formulas. With $m(\mu) = 0$ for all μ such that $V_\mu = 0$, a little juggling then permits us to rewrite (4) as follows, for *arbitrary $\mu \in \Pi(\lambda)$*:

$$(7) \qquad Tr_{V_\mu}\phi(x_\alpha)\phi(z_\alpha) = \sum_{i=0}^{\infty} m(\mu + i\alpha)\,(\mu + i\alpha, \alpha).$$

22.3. Freudenthal's formula

Let ϕ, V be as in (22.2), dim $V > 1$. Recall from (22.1) the universal Casimir element $c_L = \sum_{i=1}^{\ell} h_i k_i + \sum_{\alpha \in \Phi} x_\alpha z_\alpha$. Since ϕ is irreducible, $\phi(c_L)$ is multiplication by a nonzero scalar, say c. Fix a weight μ of V. We want to calculate $Tr_{V_\mu}\phi(c_L) = cm(\mu)$.

First of all, $\phi(h_i)$ is just scalar multiplication by $\mu(h_i)$ in V_μ, and similarly for $\phi(k_i)$. Let $t_\mu \in H$ satisfy $\mu(h) = \kappa(t_\mu, h)$ for all $h \in H$ (as in §8). Write $t_\mu = \sum_i a_i h_i$; then by definition $\mu(h_i) = \sum_j a_j \kappa(h_j, h_i)$ and $\mu(k_i) = \sum_j a_j \kappa(h_j, k_i) = a_i$ (by duality). Therefore, $(\mu, \mu) = \sum_{i,j} a_i a_j \kappa(h_j, h_i) = \sum_i \mu(h_i)\mu(k_i)$, whence:

$$(8) \qquad Tr_{V_\mu}\phi(h_i)\phi(k_i) = m(\mu)\,(\mu, \mu).$$

Combining (8) with formula (7) in (22.2), we have:

$$(9) \qquad cm(\mu) = (\mu, \mu)m(\mu) + \sum_{\alpha \in \Phi}\sum_{i=0}^{\infty} m(\mu + i\alpha)\,(\mu + i\alpha, \alpha).$$

Notice that *the terms $m(\mu)\,(\mu, \alpha)$ and $m(\mu)\,(\mu, -\alpha)$ both occur (and cancel), so we can omit the index $i = 0$.*

We claim that formula (9) remains valid for arbitrary $\mu \in \Lambda$, $\mu \notin \Pi(\lambda)$, in which case it reads: $0 = \sum_{\alpha \in \Phi}\sum_{i=1}^{\infty} m(\mu + i\alpha)\,(\mu + i\alpha, \alpha)$. Indeed, if $\mu \notin \Pi(\lambda)$, then for each $\alpha \in \Phi$ the weights (if any) of the form $\mu + i\alpha$ must occur in a string with all i positive or all i negative. In the latter case, the summand for α is 0; in the former case this is equally true, by an argument analogous to that in (22.2) for formula (6).

The preceding discussion actually shows that for each fixed $\alpha \in \Phi$ and each $\mu \in \Lambda$, we have:

$$\text{(10)} \qquad \sum_{i=-\infty}^{\infty} m(\mu + i\alpha)\,(\mu + i\alpha, \alpha) = 0.$$

In particular,

$$\text{(11)} \quad \sum_{i=1}^{\infty} m(\mu - i\alpha)\,(\mu - i\alpha, -\alpha) = m(\mu)\,(\mu, \alpha) + \sum_{i=1}^{\infty} m(\mu + i\alpha)\,(\mu + i\alpha, \alpha).$$

Substituting (11) in (9) (summation starting with $i = 1$, as remarked following (9)), we obtain finally:

$$\text{(12)} \qquad cm(\mu) = (\mu, \mu)m(\mu) + \sum_{\alpha > 0} m(\mu)\,(\mu, \alpha)$$

$$+ 2\sum_{\alpha > 0} \sum_{i=1}^{\infty} m(\mu + i\alpha)\,(\mu + i\alpha, \alpha).$$

Letting $\delta = (1/2)\sum_{\alpha > 0} \alpha$ (13.3), this can be rewritten as:

$$\text{(13)} \qquad cm(\mu) = (\mu, \mu + 2\delta)m(\mu) + 2\sum_{\alpha > 0}\sum_{i=1}^{\infty} m(\mu + i\alpha)\,(\mu + i\alpha, \alpha).$$

The only drawback to this formula is that it still involves c. But there is a special case in which we know $m(\mu)$, namely: $m(\lambda) = 1$. Moreover, $m(\lambda + i\alpha) = 0$ for all positive roots α, all $i \geq 1$. Accordingly, we can solve (13) for the value $c = (\lambda, \lambda + 2\delta) = (\lambda + \delta, \lambda + \delta) - (\delta, \delta)$. (Actually, it is not hard to compute c directly: Exercise 23.4.) These results may now be summarized in **Freudenthal's formula**.

Theorem. *Let $V = V(\lambda)$ be an irreducible L-module of highest weight λ, $\lambda \in \Lambda^+$. If $\mu \in \Lambda$, then the multiplicity $m(\mu)$ of μ in V is given recursively as follows:*

$$\text{(14)} \quad ((\lambda + \delta, \lambda + \delta) - (\mu + \delta, \mu + \delta))m(\mu) = 2\sum_{\alpha > 0}\sum_{i=1}^{\infty} m(\mu + i\alpha)\,(\mu + i\alpha, \alpha). \qquad \square$$

It still has to be observed that Freudenthal's formula provides an *effective* method for calculating multiplicities, starting with $m(\lambda) = 1$. Thanks to Proposition 21.3, Lemma C of (13.4) shows that for $\mu \in \Pi(\lambda)$, $\mu \neq \lambda$, the quantity $(\lambda + \delta, \lambda + \delta) - (\mu + \delta, \mu + \delta)$ is nonzero; so $m(\mu) = 0$ whenever this quantity is 0, $\mu \neq \lambda$. Therefore $m(\mu)$ is known provided all $m(\mu + i\alpha)$ ($i \geq 1$, $\alpha > 0$) are known, i.e., provided all $m(\nu)$, $\mu \prec \nu \prec \lambda$, are known. (Some concrete examples will be given below.)

In practice, the use of Freudenthal's formula can be made more efficient by exploiting the fact that weights conjugate under the Weyl group have the same multiplicity (Theorem 21.2). There exist computer programs for carrying out the calculations involved. Notice that the inner product can be normalized in any convenient way, since $m(\mu)$ appears as a quotient.

22.4. Examples

To use Freudenthal's formula in any given case, we have to be able to compute explicitly the bilinear form on Λ. The form used above was the "natural" one (dual to the Killing form), but it may be normalized by any convenient scalar multiplication in view of the above remarks. One popular procedure is to require that all squared root lengths be 1, 2, or 3, the smallest being 1 (for each irreducible component of Φ). Alternatively, the inner product used in the construction of root systems in §12 can be chosen.

Example 1. $L = \mathfrak{sl}(3, \mathsf{F})$, $\Phi = \{\pm\alpha_1, \pm\alpha_2, \pm(\alpha_1+\alpha_2)\}$, $\Delta = \{\alpha_1, \alpha_2\}$, $\alpha_1 = 2\lambda_1 - \lambda_2$, $\alpha_2 = -\lambda_1 + 2\lambda_2$, $\lambda_1 = (1/3)(2\alpha_1+\alpha_2)$, $\lambda_2 = (1/3)(\alpha_1+2\alpha_2)$. Require $(\alpha_i, \alpha_i) = 1$, so that $(\alpha_1, \alpha_2) = -1/2$, $(\lambda_i, \lambda_i) = 1/3$, and $(\lambda_1, \lambda_2) = 1/6$. Set $\lambda = \lambda_1 + 3\lambda_2$. Then Freudenthal's formula yields the list of multiplicities in Table 1. Other data are listed also, for the reader's convenience. Weights are grouped by "level": calculation of $m(\mu)$ requires data only from higher levels. The reader should draw a weight diagram, in the manner of Figure 1 of (21.3).

Table 1.

μ	$m(\mu)$	$(\mu+\delta, \mu+\delta)$	$\mu = m_1\lambda_1 + m_2\lambda_2$
$\{\lambda$	1	28/3	$\lambda_1 + 3\lambda_2$
$\{\lambda-\alpha_1$	1	25/3	$-\lambda_1 + 4\lambda_2$
$\{\lambda-\alpha_2$	1	19/3	$2\lambda_1 + \lambda_2$
$\{\lambda-\alpha_1-\alpha_2$	2	13/3	$2\lambda_2$
$\{\lambda-2\alpha_2$	1	16/3	$3\lambda_1 - \lambda_2$
$\{\lambda-\alpha_1-2\alpha_2$	2	7/3	λ_1
$\{\lambda-2\alpha_1-\alpha_2$	1	13/3	$-2\lambda_1 + 3\lambda_2$
$\{\lambda-3\alpha_2$	1	19/3	$4\lambda_1 - 3\lambda_2$
$\{\lambda-2\alpha_1-2\alpha_2$	2	4/3	$-\lambda_1 + \lambda_2$
$\{\lambda-\alpha_1-3\alpha_2$	2	7/3	$2\lambda_1 - 2\lambda_2$
$\{\lambda-\alpha_1-4\alpha_2$	1	13/3	$3\lambda_1 - 4\lambda_2$
$\{\lambda-2\alpha_1-3\alpha_2$	2	1/3	$-\lambda_2$
$\{\lambda-3\alpha_1-2\alpha_2$	1	7/3	$-3\lambda_1 + 2\lambda_2$
$\{\lambda-2\alpha_1-4\alpha_2$	1	4/3	$\lambda_1 - 3\lambda_2$
$\{\lambda-3\alpha_1-3\alpha_2$	2	1/3	$-2\lambda_1$
$\{\lambda-3\alpha_1-4\alpha_2$	1	1/3	$-\lambda_1 - 2\lambda_2$
$\{\lambda-4\alpha_1-3\alpha_2$	1	7/3	$-4\lambda_1 + \lambda_2$
$\{\lambda-4\alpha_1-4\alpha_2$	1	4/3	$-3\lambda_1 - \lambda_2$

Example 2. Let L be the simple algebra of type G_2. The root system of L is constructed explicitly in (12.1). Recall that α_1 is short and α_2 is long, so that $\lambda_1 = 2\alpha_1 + \alpha_2$, $\lambda_2 = 3\alpha_1 + 2\alpha_2$. Some information obtained by using Freudenthal's formula is listed in Table 2. The weight $m_1\lambda_1 + m_2\lambda_2$ is abbreviated there by $m_1 m_2$. Rows are indexed by highest weights λ, columns by dominant weights μ, and the intersection of row λ with column μ contains the integer $m_\lambda(\mu)$ (when this is nonzero). The reader should verify parts of the table for himself.

Table 2.

	00	10	01	20	11	30	02	21	40	12	31	50	03	22
00	1													
10	1	1												
01	2	1	1											
20	3	2	1	1										
11	4	4	2	2	1									
30	5	4	3	2	1	1								
02	5	3	3	2	1	1	1							
21	9	8	6	5	3	2	1	1						
40	8	7	5	5	3	2	1	1	1					
12	10	10	7	7	5	3	2	2	1	1				
31	16	14	12	10	7	6	4	3	2	1	1			
50	12	11	9	8	6	5	3	3	2	1	1	1		
03	9	7	7	5	4	4	3	2	1	1	1	0	1	
22	21	19	16	15	11	9	7	6	4	3	2	1	1	1

22.5. Formal characters

Let $\Lambda \subset H^*$ be, as before, the lattice of integral linear functions. If $V = V(\lambda)$, $\lambda \in \Lambda^+$, we want to consider a formal sum of the weights $\mu \in \Pi(\lambda)$, each μ occurring in the sum $m(\mu)$ times. However, "$\mu + \nu$" would be a poor notation to use in such a formal sum, since this already has a concrete meaning in Λ. Therefore we introduce the **group ring** of Λ over $\mathbf{Z}$, denoted $\mathbf{Z}[\Lambda]$. By definition, $\mathbf{Z}[\Lambda]$ is a free $\mathbf{Z}$-module with basis elements $e(\lambda)$ in one-one correspondence with the elements λ of Λ, with the addition denoted $e(\lambda) + e(\mu)$. $\mathbf{Z}[\Lambda]$ becomes a commutative ring if we decree that $e(\lambda)\,e(\mu) = e(\lambda + \mu)$ and extend by linearity. (There is an identity element: $e(0)$.) $\mathscr{W}$ acts naturally on $\mathbf{Z}[\Lambda]$, by permuting the $e(\lambda)$: $\sigma e(\lambda) = e(\sigma\lambda)$.

Now it makes good sense to define the **formal character** $ch_{V(\lambda)}$, or just ch_λ, of $V(\lambda)$ as the element $\sum_{\mu \in \Pi(\lambda)} m_\lambda(\mu) e(\mu)$ of $\mathbf{Z}[\Lambda]$. (Since $m_\lambda(\mu) = 0$ whenever $\mu \notin \Pi(\lambda)$, we can even extend the summation to all $\mu \in \Lambda$.) For example, if $L = \mathfrak{sl}(2, F)$, the formal character of $V(\lambda)$ is given by $ch_\lambda = e(\lambda) + e(\lambda - \alpha) + e(\lambda - 2\alpha) + \ldots + e(\lambda - m\alpha)$, $m = \langle \lambda, \alpha \rangle$. More generally, if V is an arbitrary (finite dimensional) L-module, there is an essentially unique decomposition $V = V(\lambda_1) \oplus \ldots \oplus V(\lambda_t)$, $\lambda_i \in \Lambda^+$, thanks to Weyl's Theorem and the classification theory (§21). So $ch_V = \sum_{i=1}^{t} ch_{\lambda_i}$ may be called the formal character of V. Notice that each $\sigma \in \mathscr{W}$ fixes ch_V, since σ permutes weight spaces in each irreducible summand of V (Theorem 21.2).

Knowledge of ch_V actually enables us to recover the irreducible constituents of V, because of the following result.

Proposition A. *Let $f = \sum_{\lambda \in \Lambda} c(\lambda)e(\lambda)$, $c(\lambda) \in \mathbf{Z}$, be fixed by all elements of $\mathscr{W}$. Then f can be written in one and only one way as a $\mathbf{Z}$-linear combination of the ch_λ $(\lambda \in \Lambda^+)$.*

Proof. It is clear that $f = \sum_{\lambda \in \Lambda^+} c(\lambda) (\sum_{\sigma \in \mathscr{W}} e(\sigma\lambda))$. For each $\lambda \in \Lambda^+$ such that $c(\lambda) \neq 0$, the set of dominant $\mu \prec \lambda$ is finite (Lemma B of (13.2)). Let M_f be the totality of such μ (for all such λ), so M_f is finite. Let $\lambda \in \Lambda^+$ be maximal among the $\lambda \in \Lambda^+$ for which $c(\lambda) \neq 0$, and set $f' = f - c(\lambda)ch_\lambda$, so clearly f' again satisfies the hypothesis of the proposition. We know that the dominant μ figuring in ch_λ all satisfy $\mu \prec \lambda$, so they all lie in M_f. This shows that $M_{f'} \subset M_f$. The inclusion is proper, because $\lambda \notin M_{f'}$. By induction on Card (M_f), we can write f' in the desired form; then f also has the desired form. To start the induction, notice that the case Card $(M_f) = 1$ is trivial: In this case a minimal dominant weight λ is the only dominant weight figuring in f, whence $f = c(\lambda)ch_\lambda$, where $ch_\lambda = \sum_{\sigma \in \mathscr{W}} e(\sigma\lambda)$. The uniqueness assertion is left to the reader (Exercise 8). ▯

One advantage in being able to multiply formal characters is brought out next.

Proposition B. *Let V, W be (finite dimensional) L-modules. Then $ch_{V \otimes W} = ch_V.ch_W$.*

Proof. On the one hand, from the way in which the action of L on $V \otimes W$ is defined (6.1), it is clear that the weights of $V \otimes W$ are those of the form $\lambda + \mu$ (λ a weight of V, μ of W), each occurring with multiplicity

$$\sum_{\pi + \pi' = \lambda + \mu} m_V(\pi)m_W(\pi')$$

(cf. Exercise 21.7). But this is also what we get if we formally multiply ch_V by ch_W. ▯

Exercises

1. Let $\lambda \in \Lambda^+$. Prove, without using Freudenthal's formula, that $m_\lambda(\lambda - k\alpha) = 1$ for $\alpha \in \Delta$ and $0 \leq k \leq \langle \lambda, \alpha \rangle$.
2. Prove that c_L is in the *center* of $\mathfrak{U}(L)$ (cf. (23.2)). [Imitate the calculation in (6.2), with ϕ omitted.] Show also that c_L is independent of the basis chosen for L.
3. In Example 1 (22.4), determine the $\mathscr{W}$-orbits of weights, thereby verifying directly that $\mathscr{W}$-conjugate weights have the same multiplicity (cf. Theorem 21.2). [Cf. Exercise 13.12.]
4. Verify the multiplicities shown in Figure 1 of (21.3).

5. Use Freudenthal's formula and the data for A_2 in Example 1 (22.4) to compute multiplicities for $V(\lambda)$, $\lambda = 2\lambda_1 + 2\lambda_2$. Verify in particular that dim $V(\lambda) = 27$ and that the weight 0 occurs with multiplicity 3. Draw the weight diagram.

6. For L of type G_2, use Table 2 of (22.4) to determine all weights and their multiplicities for $V(\lambda)$, $\lambda = \lambda_1 + 2\lambda_2$. Compute dim $V(\lambda) = 286$. [Cf. Exercise 13.12.]

7. Let $L = \mathfrak{sl}(2, F)$, and identify $m\lambda_1$ with the integer m. Use Propositions A and B of (22.5), along with Theorem 7.2, to derive the **Clebsch-Gordan formula**: If $n \le m$, then $V(m) \otimes V(n) \cong V(m+n) \oplus V(m+n-2) \oplus \ldots \oplus V(m-n)$, $n+1$ summands in all. (Cf. Exercise 7.6.)

8. Prove the uniqueness part of Proposition 22.5A.

Notes

The proof of Freudenthal's formula is taken from Jacobson [1]; see also Freudenthal [1] and Freudenthal–de Vries [1]. For computational aspects, cf. Agrawala, Belinfante [1], Beck, Kolman [1], Krusemeyer [1], Burgoyne, Williamson [1]. In Part II of Freudenthal [1], the case E_8 is illustrated in detail. The data in Table 2 is taken from Springer [1].

23. Characters

Our object is to prove a theorem of Harish-Chandra on "characters" associated with the infinite dimensional modules $Z(\lambda)$, $\lambda \in H^*$ (20.3). This theorem will be used in §24 to obtain a simple algebraic proof of Weyl's classical result on characters of finite dimensional modules. As a preliminary (which is also of independent interest) we shall prove in (23.1) a theorem of Chevalley on "lifting" invariants. None of this depends on Freudenthal's formula (22.3).

23.1. Invariant polynomial functions

If V is a finite dimensional vector space, the symmetric algebra $\mathfrak{S}(V^*)$ (see (17.1)) is called the algebra of **polynomial functions** on V, and is denoted $\mathfrak{P}(V)$. When a fixed basis $(f_1, \ldots, f_n)$ of V^* is given, $\mathfrak{P}(V)$ becomes identified with the algebra of polynomials in n variables $f_1, \ldots, f_n$. In this subsection we consider $\mathfrak{P}(L)$ and $\mathfrak{P}(H)$.

Since the weight lattice Λ spans H^*, the polynomials in the $\lambda \in \Lambda$ span $\mathfrak{P}(H)$. By the process of polarization (Exercise 5) the pure powers λ^k ($\lambda \in \Lambda$, $k \in \mathbf{Z}^+$) already suffice to span $\mathfrak{P}(H)$. Now consider $\mathscr{W}$, which acts on H^* and hence on $\mathfrak{P}(H)$. Let $\mathfrak{P}(H)^{\mathscr{W}}$ be the subalgebra consisting of polynomial functions fixed by all $\sigma \in \mathscr{W}$; this is the algebra of $\mathscr{W}$-**invariant polynomial** functions on H. (For example, if $L = \mathfrak{sl}(2, F)$, $\lambda = $ fundamental dominant weight, then $\mathfrak{P}(H)^{\mathscr{W}}$ is the algebra with 1 generated by λ^2.) If we write $\text{Sym} f$ for the sum of all distinct $\mathscr{W}$-conjugates of $f \in \mathfrak{P}(H)$, then it is clear that the collection of all $\text{Sym} \lambda^k$ ($\lambda \in \Lambda^+$, $k \in \mathbf{Z}^+$) spans $\mathfrak{P}(H)^{\mathscr{W}}$, because each $\lambda \in \Lambda$ is $\mathscr{W}$-conjugate to a dominant integral linear function (Lemma 13.2A).

Next let $G = \text{Int } L$, which is generated by all $\exp \text{ad } x$ (x nilpotent). Then G acts naturally on $\mathfrak{P}(L)$, via $(\sigma f)(x) = f(\sigma^{-1}x)$ ($\sigma \in G$, $f \in \mathfrak{P}(L)$), and we denote the fixed elements of $\mathfrak{P}(L)$ by $\mathfrak{P}(L)^G$. These are the **G-invariant polynomial functions** on L.

Many examples of G-invariant polynomial functions can be constructed via representation theory, as follows. Let $\phi: L \to \mathfrak{gl}(V)$ be an irreducible (finite dimensional) representation of L, of highest weight $\lambda \in \Lambda^+$, and let $z \in N = \coprod_{\alpha \succ 0} L_\alpha$, $\sigma = \exp \text{ad } z$. Define a new representation $\phi^\sigma: L \to \mathfrak{gl}(V)$ by the rule $\phi^\sigma(x) = \phi(\sigma(x))$, $x \in L$. (Check that this actually satisfies $\phi^\sigma([xy]) = [\phi^\sigma x, \phi^\sigma y]$.) Obviously ϕ^σ is again irreducible. If $v^+ \in V$ is a maximal vector, and β is any positive root, then $\phi^\sigma(x_\beta)(v^+) = \phi((1 + \text{ad } z + \frac{(\text{ad } z)^2}{2!} + \ldots)$
$(x_\beta))(v^+) = 0$, since the element of L in parentheses is still in N. Moreover, $\phi^\sigma(h)(v^+) = \phi(h + [zh])(v^+) = \phi(h)(v^+) = \lambda(h)v^+$, since $[zh] \in N$ and $\phi(N)$ $(v^+) = 0$. In other words, v^+ is again a maximal vector of weight λ for the new representation, so the two representations ϕ and ϕ^σ are *equivalent* (i.e., the two L-module structures on V are isomorphic (20.3)). Let $\psi_\sigma: V \to V$ be an L-module isomorphism, so that $\psi_\sigma(\phi(x)(v)) = \phi^\sigma(x)(\psi_\sigma(v))$ for all $v \in V$. Concretely, ψ_σ is just a change of basis in V, and this equation shows that the matrices of $\phi(x)$ and $\phi^\sigma(x) = \phi(\sigma x)$ (relative to a fixed basis of V) are *similar*. In particular, they have the same trace. If $k \in \mathbf{Z}^+$, it follows that the function $x \mapsto Tr(\phi(x)^k)$ is σ-invariant. But this is a polynomial function: starting with the (linear) coordinate functions for $\phi(x)$, the entries of $\phi(x)^k$ become polynomials in these, and the trace is a linear combination of such polynomials. Notice too that the invariance of the trace function is independent of the original choice of base (or positive roots) and even the choice of H, so that $x \mapsto Tr(\phi(x)^k)$ is in fact fixed by all generators of G (cf. Exercise 16.2), hence by G itself.

Now we are ready to compare $\mathfrak{P}(L)^G$ with $\mathfrak{P}(H)^{\mathscr{W}}$ (this being the whole point of the discussion). Any polynomial function f on L, when restricted to H, is a polynomial function on H: this is obvious if a basis of H is extended to a basis of L and f is written as a polynomial in the elements of the dual basis. If f happens to be G-invariant, then in particular it is fixed by each of the inner automorphisms τ_α ($\alpha \in \Phi$) constructed in (14.3). But $\tau_\alpha|_H$ is the reflection σ_α, and the σ_α generate $\mathscr{W}$, so we see that $f|_H \in \mathfrak{P}(H)^{\mathscr{W}}$. We therefore obtain an algebra homomorphism $\theta: \mathfrak{P}(L)^G \to \mathfrak{P}(H)^{\mathscr{W}}$.

Theorem (Chevalley). θ *is surjective.*

Proof (*Steinberg*). By previous remarks, it will suffice to show that each $\text{Sym} \lambda^k$ ($\lambda \in \Lambda^+$, $k \in \mathbf{Z}^+$) lies in the image of θ. For this we use upward induction on the partial ordering of Λ^+, starting with λ minimal (possibly 0). (Recall from Lemma 13.2B that the dominant weights lying below a given one are finite in number.) Since λ is minimal, no other $\mu \in \Lambda^+$ can occur as a weight of the irreducible representation ϕ whose highest weight is λ. In view of Theorems 20.2, 21.2, the sole weights of ϕ are the $\mathcal{W}$-conjugates of λ, each of multiplicity one. Now $x \mapsto Tr (\phi(x)^k)$ is a G-invariant polynomial function f, whose restriction to H is Sym λ^k. So Sym $\lambda^k = \theta(f)$.

For the induction step, fix $\lambda \in \Lambda^+$, $k \in \mathbf{Z}^+$. Let ϕ again denote the irreducible representation of highest weight λ, f the function $x \mapsto Tr(\phi(x)^k)$. Then $f|_H = $ Sym $\lambda^k + \sum m_\lambda(\mu)$ Sym μ^k (Theorem 21.2), where we sum over $\mu \precneqq \lambda$, $\mu \in \Lambda^+$. The terms involving $\mu \precneqq \lambda$ are all liftable to $\mathfrak{P}(L)^G$, by induction, so finally Sym λ^k is liftable. $\square$

Let us make one further observation about $\mathfrak{P}(L)^G$. Call a polynomial function $x \mapsto Tr(\phi(x)^k)$ as above a **trace polynomial**. If $x = x_s + x_n$ is the Jordan decomposition of x, then $\phi(x) = \phi(x_s) + \phi(x_n)$ is the (usual) Jordan decomposition of $\phi(x)$ (cf. (6.4)). Since $\phi(x_s)$ and $\phi(x_n)$ commute, all terms except $\phi(x_s)^k$ in the expansion of $(\phi(x_s) + \phi(x_n))^k$ are nilpotent, hence of trace 0. Therefore, a trace polynomial is completely determined by its values at semisimple elements of L. The proof of Chevalley's theorem actually shows that θ maps the subalgebra $\mathfrak{T} \subset \mathfrak{P}(L)^G$ generated by trace polynomials onto $\mathfrak{P}(H)^{\mathcal{W}}$. In fact, $\theta|_{\mathfrak{T}}$ is *injective* as well as surjective: $\theta(f) = 0$ means that $f|_H = 0$. Each semisimple element of L lies in some maximal toral subalgebra, hence (16.4) is conjugate under G to an element of H. Therefore, f vanishes on all semisimple elements of L, forcing $f = 0$ (by the above remarks).

Using some elementary algebraic geometry (see Appendix below), it can be shown directly that θ *is injective*; as a corollary, $\mathfrak{P}(L)^G$ *is generated by trace polynomials*. (We shall not need these results, however.) In (23.3) we shall allow ourselves to write θ^{-1}, but the reader can easily check that the argument does not depend essentially on the injectivity of θ.

23.2. Standard cyclic modules and characters

Let $\mathfrak{Z}$ be the **center** of $\mathfrak{U}(L)$, i.e., the set of elements commuting with all $x \in \mathfrak{U}(L)$, or equivalently, with all $x \in L$. An automorphism $\sigma: L \to L$ extends uniquely to an automorphism of $\mathfrak{U}(L)$, so in particular $G = $ Int L acts on $\mathfrak{U}(L)$, mapping $\mathfrak{Z}$ onto itself. The following fact will be needed in (23.3).

Lemma. $\mathfrak{Z}$ *is precisely the set of G-invariants of* $\mathfrak{U}(L)$.

Proof. On the one hand, $\mathfrak{Z}$ commutes with all nilpotent $x \in L$, so $0 = [xz] = $ ad x (z) $(z \in \mathfrak{Z})$, and exp ad x $(z) = z$. This implies that all $\sigma \in G$ fix z. Conversely, let G fix an element x of $\mathfrak{U}(L)$. Fix a root $\alpha \in \Phi$ and take $0 \neq x_\alpha \in L_\alpha$. If $n = $ ad x_α, suppose $n^t \neq 0$, while $n^{t+1} = 0$. Then choose

$t+1$ distinct scalars $a_1, \ldots, a_{t+1}$ in F (possible since F is infinite). By hypothesis, $1 + a_i n + (a_i^2/2!)n^2 + \ldots + (a_i^t/t!)n^t$ fixes x $(1 \le i \le t+1)$. The determinant

$$\begin{vmatrix} 1 & a_1 & a_1^2/2! & \cdots & a_1^t/t! \\ \vdots & & & & \vdots \\ 1 & a_{t+1} & a_{t+1}^2/2! & \cdots & a_{t+1}^t/t! \end{vmatrix}$$

is $(2! \; 3! \ldots t!)^{-1}$ times the Vandermonde determinant $\prod_{i<j} (a_i - a_j) \ne 0$.
Therefore, we can find scalars $b_1, \ldots, b_{t+1}$ (not all 0) satisfying: $n = \sum_{i=1}^{t+1} b_i$
(exp $a_i n$). (Strictly speaking, this is done in the space of endomorphisms of the (finite dimensional) L-submodule of $\mathfrak{U}(L)$ generated by x, cf. Exercise 17.3.) In particular, ad $x_\alpha(x) = \sum b_i$ exp (ad $a_i x_\alpha$) $(x) = (\sum b_i)x$. Since ad x_α is nilpotent, we conclude that $\sum b_i = 0$, $[x_\alpha, x] = 0$. But the x_α generate L, so x centralizes L and $x \in \mathfrak{Z}$ as required. $\square$

We remark that *the universal Casimir element c_L (22.1) belongs to $\mathfrak{Z}$*: Just imitate the calculation in (6.2), omitting mention there of ϕ.

Next we ask how $\mathfrak{Z}$ acts on the infinite dimensional module $Z(\lambda)$, $\lambda \in H^*$, which was constructed in (20.3). If v^+ is a maximal vector of $Z(\lambda)$, and $z \in \mathfrak{Z}$, notice that $h.z.v^+ = z.h.v^+ = \lambda(h)z.v^+$ $(h \in H)$, while $x_\alpha.z.v^+ = z.x_\alpha.v^+ = 0$ $(x_\alpha \in L_\alpha, \alpha \succ 0)$. Therefore, $z.v^+$ is another maximal vector of weight λ; according to Theorem 20.2, $z.v^+$ must be a scalar multiple of v^+, say $\chi_\lambda(z)v^+$. The resulting function $\chi_\lambda \colon \mathfrak{Z} \to$ F is an F-algebra homomorphism called the **character** determined by λ.

It is clear that the set of all vectors in $Z(\lambda)$ on which $z \in \mathfrak{Z}$ acts as scalar multiplication by $\chi_\lambda(z)$ is $\mathfrak{U}(L)$-stable and includes v^+, hence must be all of $Z(\lambda)$. Therefore, the action of $z \in \mathfrak{Z}$ on any submodule of $Z(\lambda)$ is scalar multiplication by $\chi_\lambda(z)$ (similarly, on any homomorphic image of $Z(\lambda)$).

It turns out that not all characters χ_λ $(\lambda \in H^*)$ are distinct. To get precise conditions for equality of characters, we define $\lambda, \mu \in H^*$ to be **linked** (written $\lambda \sim \mu$) if $\lambda + \delta$ and $\mu + \delta$ are $\mathscr{W}$-conjugate (where $\delta =$ half-sum of positive roots, as in (13.3)). It is clear that linkage is an equivalence relation. Here we shall only be concerned with integral linear functions (i.e., elements of Λ). Choose $x_\alpha \in L_\alpha$ $(\alpha \succ 0)$, $y_\alpha \in L_{-\alpha}$, so that $[x_\alpha y_\alpha] = h_\alpha$.

Proposition. *Let $\lambda \in \Lambda$, $\alpha \in \Delta$. If the integer $m = \langle \lambda, \alpha \rangle$ is nonnegative, then the coset of y_α^{m+1} in $Z(\lambda)$ is a maximal vector of weight $\lambda - (m+1)\alpha$.*

Proof. Use the formulas of Lemma 21.2, along with the fact that $h_\alpha - \langle \lambda, \alpha \rangle \cdot 1 \in I(\lambda)$ (20.3). $\square$

Corollary. *Let $\lambda \in \Lambda$, $\alpha \in \Delta$, $\mu = \sigma_\alpha(\lambda + \delta) - \delta$. Then $\chi_\lambda = \chi_\mu$.*

Proof. Since σ_α permutes the positive roots other than α and sends α to $-\alpha$ (Lemma 10.2B), $\sigma_\alpha \delta - \delta = -\alpha$. Therefore, $\mu = \sigma_\alpha(\lambda + \delta) - \delta = \sigma_\alpha \lambda - \alpha = \lambda - (\langle \lambda, \alpha \rangle + 1)\alpha$. By assumption, $\langle \lambda, \alpha \rangle \in \mathbf{Z}$. If this number is nonnegative, the proposition shows that $Z(\lambda)$ contains a homomorphic image of $Z(\mu)$ (different from 0), whence $\chi_\lambda = \chi_\mu$ by earlier remarks. If $\langle \lambda, \alpha \rangle$ is negative, then $\langle \mu, \alpha \rangle = \langle \lambda, \alpha \rangle - 2(\langle \lambda, \alpha \rangle + 1) = -\langle \lambda, \alpha \rangle - 2$ is nonnegative (unless $\langle \lambda, \alpha \rangle = -1$, in which case $\mu = \lambda$ and there is nothing to prove). So the proposition applies, with μ in place of λ, and again $\chi_\lambda = \chi_\mu$. $\square$

The corollary shows that two integral linear functions linked by a simple reflection yield the same character. Using the transitivity of linkage and the fact that $\mathscr{W}$ is generated by simple reflections (Theorem 10.3(d)), we obtain a strengthened version.

Corollary'. *Let* $\lambda, \mu \in \Lambda$. *If* $\lambda \sim \mu$, *then* $\chi_\lambda = \chi_\mu$. $\square$

This is the easy half of Harish-Chandra's Theorem (see (23.3)). We shall see below how to extend it to cover all $\lambda, \mu \in H^*$, but only the integral case is actually needed in this book.

23.3. Harish-Chandra's Theorem

Theorem (Harish-Chandra). *Let* $\lambda, \mu \in H^*$. *If* $\chi_\lambda = \chi_\mu$, *then* $\lambda \sim \mu$.

This subsection will be occupied with the proof of the theorem. The idea of the proof is not really very difficult, but there are a number of maps to keep straight. To begin with, fix a convenient basis of L, say $\{h_i, 1 \le i \le \ell;$ $x_\alpha, y_\alpha, \alpha \succ 0\}$, where $h_i = h_{\alpha_i}$, $\Delta = \{\alpha_1, \ldots, \alpha_\ell\}$. Construct PBW bases for $\mathfrak{U}(L)$ and $\mathfrak{U}(H)$ accordingly, relative to an ordering which puts the y_α first, then the h_i, then the x_α. Define a linear map $\xi : \mathfrak{U}(L) \to \mathfrak{U}(H)$ by sending each basis monomial in $h_1, \ldots, h_\ell$ to itself and all other basis elements to 0.

If v^+ is a maximal vector of the irreducible module $V(\lambda)$, $\lambda \in H^*$, consider how $z \in \mathfrak{Z}$ (expressed in terms of the above PBW basis) acts on v^+. A monomial $\prod_{\alpha \succ 0} y_\alpha^{i_\alpha} \prod_i h_i^{k_i} \prod_{\alpha \succ 0} x_\alpha^{j_\alpha}$ for which some $j_\alpha > 0$ kills v^+, while one for which all $j_\alpha = 0$ but some $i_\alpha > 0$ sends v^+ first to a multiple of itself and then to a lower weight vector. Accordingly, the only monomials which contribute to the eigenvalue $\chi_\lambda(z)$ are those for which all $i_\alpha = 0 = j_\alpha$. From this it follows at once that, with ξ as above:

$$(*) \qquad\qquad \chi_\lambda(z) = \lambda(\xi(z)), \quad z \in \mathfrak{Z}.$$

($\lambda : H \to \mathsf{F}$ extends canonically to a homomorphism of associative algebras $\mathfrak{U}(H) \to \mathsf{F}$.) Notice that the restriction of ξ to $\mathfrak{Z}$ is an *algebra* homomorphism, thanks to $(*)$.

Somehow δ must be gotten into the picture. This happens as follows. Send each h_i to $h_i - 1$, and extend linearly to a map $H \to \mathfrak{U}(H)$. This is a Lie algebra homomorphism (all Lie products being 0), so it extends to a homomorphism $\eta : \mathfrak{U}(H) \to \mathfrak{U}(H)$. Clearly η is an automorphism (with inverse sending h_i to $h_i + 1$). Let $\psi : \mathfrak{Z} \to \mathfrak{U}(H)$ be the composite homomorphism $\eta \circ \xi|_{\mathfrak{Z}}$. Recall (13.3) that $\delta = \sum_{i=1}^{\ell} \lambda_i$ (λ_i fundamental dominant weights in Λ), so $\delta(h_i) = 1$. It follows that $(\lambda + \delta)$ $(h_i - 1) = (\lambda + \delta)$ $(h_i) - (\lambda + \delta)$ $(1) = (\lambda(h_i) + 1) - 1 = \lambda(h_i)$. Therefore:

$$(**) \qquad\qquad (\lambda + \delta)(\psi(z)) = \lambda(\xi(z)) \qquad (z \in \mathfrak{Z}, \lambda \in H^*).$$

Combined with (*), this says that $\chi_\lambda(z) = (\lambda + \delta) \ (\psi(z))$. Now let λ be *integral*. By Corollary' in (23.2), all the conjugates $\sigma(\lambda + \delta)$ agree at $\psi(z)$; equivalently, $\mu = \lambda + \delta$ takes the same value at all $\mathscr{W}$-conjugates of $\psi(z)$. This being true for all $\lambda \in \Lambda$, hence for all $\mu \in \Lambda$, it follows that all linear functions take the same value at all $\mathscr{W}$-conjugates of $\psi(z)$. But then $\mathscr{W}$ must fix $\psi(z)$, $z \in \mathfrak{Z}$. We may replace $\mathfrak{U}(H)$ here by the symmetric algebra $\mathfrak{S}(H)$, because H is abelian (Example 17.2). Therefore, our conclusion is that ψ maps $\mathfrak{Z}$ into $\mathfrak{S}(H)^{\mathscr{W}}$ (the elements of $\mathfrak{S}(H)$ fixed by $\mathscr{W}$).

It has been shown that, for all $\lambda \in H^*$, $\chi_\lambda(z) = (\lambda + \delta) \ (\psi(z))$, $z \in \mathfrak{Z}$. Moreover, $\psi(z)$ is $\mathscr{W}$-invariant, so the right side does not change if we write $\sigma(\lambda + \delta)$ in place of $\lambda + \delta$. Therefore, $\chi_\lambda(z) = \chi_\mu(z)$ if $\lambda \sim \mu$ (μ linked to λ by σ). This shows that *Corollary' in (23.2) extends to all λ, $\mu \in H^*$, as remarked there.*

We just saw that $\chi_\lambda(z) = (\lambda + \delta) \ (\psi(z))$, $\lambda \in H^*$. Suppose $\chi_\lambda = \chi_\mu$. Then $\lambda + \delta$ and $\mu + \delta$ agree on $\psi(\mathfrak{Z})$, which lies in $\mathfrak{S}(H)^{\mathscr{W}}$. To prove Harish-Chandra's Theorem, we must show that $\lambda + \delta$ and $\mu + \delta$ are conjugate under $\mathscr{W}$. For this *it will suffice to prove that* $\psi(\mathfrak{Z}) = \mathfrak{S}(H)^{\mathscr{W}}$, in view of the following lemma.

Lemma. *Let* λ_1, $\lambda_2 \in H^*$ *lie in distinct $\mathscr{W}$-orbits. Then* λ_1, λ_2 *take distinct values at some element of* $\mathfrak{S}(H)^{\mathscr{W}}$ $(= \mathfrak{P}(H^*)^{\mathscr{W}})$.

Proof. This is elementary, requiring only the finiteness of $\mathscr{W}$. Begin by choosing some polynomial in $\mathfrak{S}(H)$ at which λ_1 does not vanish, but at which all other $\mathscr{W}$-conjugates of λ_1, as well as all $\mathscr{W}$-conjugates of λ_2, vanish. (Why does such a polynomial exist?) Add up the images of this polynomial, to get an element of $\mathfrak{S}(H)^{\mathscr{W}}$ at which λ_2 vanishes but λ_1 does not. ☐

The remaining task is to prove that ψ maps $\mathfrak{Z}$ onto $\mathfrak{S}(H)^{\mathscr{W}}$. There is one further map to introduce. Recall that $\mathfrak{S}(L)$ may be identified with the space of *symmetric tensors* in $\mathfrak{T}(L)$, which is complementary to the kernel J of the canonical map $\pi \colon \mathfrak{T}(L) \to \mathfrak{U}(L)$ (Corollary E of Theorem 17.3). Let $G = \mathrm{Int}\ L$, as in (23.1); G acts on $\mathfrak{T}(L)$. It is obvious that $\mathfrak{S}(L)$, J are stable under the action of G, so the linear isomorphism $\pi \colon \mathfrak{S}(L) \to \mathfrak{U}(L)$ is actually an isomorphism of G-modules. Denote by $\mathfrak{S}(L)^G$ the subspace (actually, subalgebra) of elements fixed by G. According to Lemma 23.2, π maps $\mathfrak{S}(L)^G$ onto $\mathfrak{Z} = \mathfrak{U}(L)^G$. (*Caution:* π is not an algebra homomorphism, only a linear map.)

We now have the picture: $\mathfrak{S}(L)^G \overset{\pi}{\to} \mathfrak{Z} \overset{\psi}{\to} \mathfrak{S}(H)^{\mathscr{W}}$. This bears a striking resemblance to the set-up studied in (23.1): $\mathfrak{P}(L)^G \overset{\theta}{\to} \mathfrak{P}(H)^{\mathscr{W}}$. Indeed, we can even identify (canonically) L with L^*, H with H^*, by means of the Killing form (which is nondegenerate on both L and H), and the actions of G, $\mathscr{W}$ are compatible with these identifications. Consider the resulting diagram:

$$
\begin{array}{ccccc}
\mathfrak{S}(L)^G & \overset{\pi}{\longrightarrow} & \mathfrak{Z} & \overset{\psi}{\longrightarrow} & \mathfrak{S}(H)^{\mathscr{W}} \\
\uparrow & & & & \uparrow \\
\mathfrak{P}(L)^G & & \underset{\theta}{\longrightarrow} & & \mathfrak{P}(H)^{\mathscr{W}}
\end{array}
$$

Unfortunately, this diagram is not quite commutative. To see what is going on, we pause to look at a simple example.

Example. Let $L = \mathfrak{sl}(2, \text{F})$, with standard basis (x, y, h). The dual basis (x^*, y^*, h^*) may be identified with $(\frac{1}{4}y, \frac{1}{4}x, \frac{1}{8}h)$ via the Killing form (Exercise 5.5). If λ is the fundamental dominant weight ($\lambda = \frac{1}{2}\alpha$), then λ identifies here with h^*, and λ^2 generates $\mathfrak{P}(H)^{\mathscr{W}}$. λ being the highest weight of the usual representation of L, an easy calculation (Exercise 1) shows that the trace polynomial $h^{*2} + x^*y^*$ equals $\theta^{-1}(\lambda^2)$. Under the identification of $\mathfrak{P}(L)^G$ with $\mathfrak{S}(L)^G$, this becomes the symmetric tensor $(1/64) (h \otimes h) + (1/32) (x \otimes y + y \otimes x)$. π maps this element to $(1/64)h^2 + (1/32)xy + (1/32)yx \in \mathfrak{Z}$. In turn, to calculate the image under ψ, we must rewrite this element in the PBW basis (relative to the ordering y, h, x) as $(1/64)h^2 + (2/32)yx + (1/32)h$. ξ sends this to $(1/64) (h^2 + 2h)$, and η in turn to the $\mathscr{W}$-invariant $(1/64)$ $(h^2 - 1)$. Reverting to $\mathfrak{P}(H)^{\mathscr{W}}$, this yields $\lambda^2 - 1/64$. Therefore, the diagram does fail to commute. Nevertheless, the discrepancy is measured by an invariant (here the scalar $1/64$) of lower "degree" than the element we started with.

This example suggests how to complete the proof that ψ is surjective (for $\mathfrak{sl}(2, \text{F})$, it is the proof!). First of all, we agree to identify $\mathfrak{P}(L)^G$ with $\mathfrak{S}(L)^G$ and $\mathfrak{P}(H)^{\mathscr{W}}$ with $\mathfrak{S}(H)^{\mathscr{W}}$. Next, it is obvious that if a polynomial in $\mathfrak{S}(H)$ is fixed by $\mathscr{W}$, then so are its homogeneous parts. Therefore, it will suffice to lift *homogeneous* polynomials to $\mathfrak{Z}$; in particular, we can use induction on degree (constants being trivially liftable).

The maps θ and $\xi \circ \pi$ are now essentially the same (recall how each is defined), except that we have to rewrite the "symmetric" element $\pi(f)$ in the PBW basis before applying ξ. This introduces some new terms. However, if f has degree k, then $\pi(f)$ is a sum of terms $x_1 \ldots x_k$ in $\mathfrak{U}(L)$ (x_i among the fixed basis elements of L), so the new terms obtained by commutating clearly have the form $x_1 \ldots x_j$ for $j < k$.

The map η (sending h_i to $h_i - 1$) has no effect on highest degree terms of polynomials in $\mathfrak{S}(H)$. The upshot is that we recover our original homogeneous element of $\mathfrak{S}(H)^{\mathscr{W}}$, modulo terms of lower degree, when we go around the diagram using θ^{-1}, then π, then ψ. The lower degree terms are, by induction, images under ψ of elements of $\mathfrak{Z}$, so the argument is complete. $\square$

Appendix

One fact was left unproved in this section: that *the restriction map* $\mathfrak{P}(L)^G \to \mathfrak{P}(H)^{\mathscr{W}}$ *is injective* (23.1). This fact is inessential to the proof of Harish-Chandra's Theorem, but it would be less than satisfactory to pass over it in silence. It can be formulated as a simple density argument within the framework of (affine) algebraic geometry.

Let $A = \text{F}^n$ (called **affine n-space**); we ignore the vector space structure of A here. Let $\text{F}[T] = \text{F}[T_1, \ldots, T_n]$ be the polynomial ring in n indeterminates.

If I is an ideal in $F[T]$, let $\mathscr{V}(I) = \{x = (x_1, \ldots, x_n) \in A | f(x) = 0$ for all $f \in I\}$. We topologize A by declaring the sets $\mathscr{V}(I)$ to be **closed**; obviously $\varnothing$ and A are closed, while the fact that finite unions or arbitrary intersections of closed sets are closed is easy to verify (using, e.g., $\mathscr{V}(\sum_\alpha I_\alpha) = \bigcap_\alpha \mathscr{V}(I_\alpha)$).

This topology on A is called the **Zariski topology**. (In case $F = \mathbf{R}$ or $\mathbf{C}$, a Zariski-closed set is also closed in the usual topology on F^n, but not conversely.)

If $f(T) \in F[T]$, the function $x \mapsto f(x)$ ($A \to F$) is called a **polynomial function** on A. Evidently such a function is continuous in the Zariski topology, F being given the Zariski topology of affine 1-space. Since F is infinite, the only polynomial vanishing on A is the zero polynomial (as is well known).

A subset of A is called **irreducible** if it cannot be covered by two closed sets neither of which already covers it. ("Irreducible" implies "connected", but not conversely.)

Lemma. A *is irreducible.*

Proof. Let $A = \mathscr{V}(I_1) \cup \mathscr{V}(I_2)$, and suppose both of these closed sets are proper. Then $I_1 \neq 0$, $I_2 \neq 0$. Let $f \in I_1$, $g \in I_2$ be nonzero polynomials. Then $fg \neq 0$, but fg vanishes identically on A, which is absurd. ⬜

Corollary. *Any nonempty open set in* A *is dense.*

Proof. Let U be a nonempty open set. If U is not dense, then there exists a nonempty open set V in A with $U \cap V = \varnothing$; the closed sets $A - U$, $A - V$ are then proper, and cover A, contradicting the lemma. ⬜

Return now to the situation of §23: L semisimple, H a CSA (etc.). Fix a basis of L, so that L becomes identified with affine n-space ($n = \dim L$) and $\mathfrak{P}(L)$ with the polynomial functions in the above sense. Relative to this basis, ad x is represented by an $n \times n$ matrix, whose coordinates are linear (hence polynomial) functions on L. In turn, let T be an indeterminate, and write (for $x \in L$) $p_x(T) = $ characteristic polynomial of ad $x = \sum_{i=0}^n c_i(x)T^i$. It is clear that each c_i is a polynomial function on L.

Call the ρ-**rank** of L the smallest integer m for which c_m is not identically 0, and call $x \in L$ ρ-**regular** if $c_m(x) \neq 0$. In words, x is ρ-regular if and only if 0 has the smallest possible multiplicity as an eigenvalue of ad x. This shows that x is ρ-regular if and only if x_s is (since they have the same characteristic polynomial); in particular, ρ-regular semisimple elements exist. Obviously the set $\mathscr{R}$ of all ρ-regular elements of L is open; being open and nonempty, it is *dense* (by the above corollary).

If $x \in L$ is semisimple, x lies in a maximal toral subalgebra, so the conjugacy theorem implies that x is conjugate under G to some element of H. But if $h \in H$, we know that $\dim C_L(h) \geq \ell = \text{rank } L$; we also know that H possesses elements (called *regular*) for which $\dim C_L(h) = \ell$ (15.3). Since ρ-regular semisimple elements exist, they must therefore coincide with the regular semisimple elements (and $m = \ell$). But no nilpotent element other than 0 centralizes a regular semisimple element. In view of the preceding paragraph, x ρ-regular implies $x = x_s$. *This allows us to describe $\mathcal{R}$ as the set of all regular semisimple elements.*

Now let $f \in \mathfrak{P}(L)^G$, $f|_H = 0$. This implies in particular that f vanishes on $\mathcal{R}$, which is dense, so $f = 0$. Therefore $\theta: \mathfrak{P}(L)^G \to \mathfrak{P}(H)^{\mathscr{W}}$ is injective.

Exercises

1. In the Example in (23.3), verify that the trace polynomial is given correctly.

2. For the algebras of type A_2, B_2, G_2, compute explicit generators for $\mathfrak{P}(H)^{\mathscr{W}}$ in terms of the fundamental dominant weights λ_1, λ_2. Show how some of these lift to $\mathfrak{Z}$, using the algorithm of this section. (Notice too that in each case $\mathfrak{P}(H)^{\mathscr{W}}$ is a polynomial algebra with $\ell = 2$ generators.)

3. Show that Proposition 23.2 remains valid when λ is an arbitrary linear function on H, provided only that $\langle \lambda, \alpha \rangle$ is an integer.

4. From the formula (*) $\chi_\lambda(z) = \lambda(\xi(z))$ of (23.3), compute directly the value of the universal Casimir element c_L (22.1) on $V(\lambda)$, $\lambda \in \Lambda^+ : (\lambda+\delta, \lambda+\delta) - (\delta, \delta)$. [Recall how t_α and h_α, resp. z_α and y_α, are related. Rewrite c_L in the ordering of a PBW basis, and use the fact derived in (22.3) that $(\mu, \mu) = \sum_i \mu(h_i)\mu(k_i)$ for any weight μ.]

5. Prove that any polynomial in n variables over F (char F $= 0$) is a linear combination of powers of linear polynomials. [Use induction on n. Expand $(T_1 + aT_2)^k$ and then use a Vandermonde determinant argument to show that kth powers of linear polynomials span a space of correct dimension when $n = 2$.]

6. If $\lambda \in \Lambda^+$ prove that all μ linked to λ satisfy $\mu \prec \lambda$, hence that all such μ occur as weights of $Z(\lambda)$.

7. Let $\mathfrak{D} = [\mathfrak{U}(L), \mathfrak{U}(L)]$ be the subspace of $\mathfrak{U}(L)$ spanned by all $xy - yx$ ($x, y \in \mathfrak{U}(L)$). Prove that $\mathfrak{U}(L)$ is the direct sum of the subspaces $\mathfrak{D}$ and $\mathfrak{Z}$ (thereby allowing one to extend χ_λ to all of $\mathfrak{U}(L)$ by requiring it to be 0 on $\mathfrak{D}$). [Recall from Exercise 17.3 that $\mathfrak{U}(L)$ is the sum of finite dimensional L-modules, hence is completely reducible because L is semisimple. Show that $\mathfrak{Z}$ is the sum of all trivial L-submodules of $\mathfrak{U}(L)$, while $\mathfrak{D}$ coincides with the space of all ad $x(y)$, $x \in L$, $y \in \mathfrak{U}(L)$, the latter being complementary to $\mathfrak{Z}$.]

8. Prove that the weight lattice Λ is Zariski dense in H^* (see Appendix), H^* being identified with affine ℓ-space. Use this to give another proof that Corollary' in (23.2) extends to all λ, $\mu \in H^*$.

9. Every F-algebra homomorphism $\chi : \mathfrak{Z} \to$ F is of the form χ_λ for some $\lambda \in H^*$. [View χ as a homomorphism $\mathfrak{S}(H)^{\mathscr{W}} \to$ F and show that its kernel generates a proper ideal in $\mathfrak{S}(H)$.]

10. Prove that the map $\psi : \mathfrak{Z} \to \mathfrak{S}(H)^{\mathscr{W}}$ is independent of the choice of Δ.

Notes

Steinberg's proof of Chevalley's theorem on invariant polynomials is written down in an appendix to the thesis of Verma [1], and in Varadarajan [1], §5.6. For accounts of Harish-Chandra's work on "characters", see Harish-Chandra [1], Séminaire "Sophus Lie" [1], Exposé 19, as well as Varadarajan [1], §5.7. We have followed Verma [1].

24. Formulas of Weyl, Kostant, and Steinberg

The notation is that of §23. We are going to use Harish-Chandra's Theorem (23.3) to obtain several remarkable formulas for the characters and multiplicities of finite dimensional L-modules. No use is made of Freudenthal's formula (22.3).

24.1. Some functions on H*

For $\lambda \in \Lambda^+$, the *formal character* ch_λ of $V(\lambda)$ was introduced in (22.5): $ch_\lambda = \sum_{\mu \in \Lambda} m_\lambda(\mu)e(\mu)$, where the $e(\mu)$ form a basis for the group ring $\mathbf{Z}[\Lambda]$. It is also convenient to define formal characters for the infinite dimensional modules $Z(\lambda)$, but here the infinite formal sums would be awkward to manipulate. Instead, we use a more suggestive formalism. $\mathbf{Z}[\Lambda]$ can be viewed as the set of $\mathbf{Z}$-valued functions on Λ (0 outside a finite set); the reader can easily check that the product operation becomes **convolution**, $f * g(\lambda) = \sum_{\mu + \nu = \lambda} f(\mu)g(\nu)$.

Let $\mathfrak{X}$ be the space of all F-valued functions f on H^* whose **support** (defined to be the set of $\lambda \in H^*$ for which $f(\lambda) \neq 0$) is included in a finite union of sets of the form $\{\lambda - \sum_{\alpha > 0} k_\alpha \alpha, k_\alpha \in \mathbf{Z}^+\}$. (Such a set is of course the set of weights occurring in a module $Z(\lambda)$, $\lambda \in H^*$.)

A moment's thought shows that $\mathfrak{X}$ is closed under convolution; thus it becomes a commutative, associative F-algebra, containing the formal character ch_V of any standard cyclic L-module.

The reader may find it convenient at times to think of $f \in \mathfrak{X}$ as a formal combination (with F-coefficients) of the $\lambda \in H^*$. What corresponds to our earlier $e(\lambda)$? Clearly this is just the *characteristic function* $\varepsilon_\lambda(\lambda) = 1$, $\varepsilon_\lambda(\mu) = 0$ if $\mu \neq \lambda$. Notice that ε_0 is the identity element of the ring $\mathfrak{X}$, and that $\varepsilon_\lambda * \varepsilon_\mu = \varepsilon_{\lambda+\mu}$. The Weyl group $\mathscr{W}$ acts on $\mathfrak{X}$ by $(\sigma^{-1}f)(\lambda) = f(\sigma\lambda)$; in particular, $\sigma^{-1}(\varepsilon_\lambda) = \varepsilon_{\sigma\lambda}$.

Two other useful elements of $\mathfrak{X}$ must now be introduced. First, let $p(\lambda)$ be the number of sets of nonnegative integers $\{k_\alpha, \alpha > 0\}$ for which $-\lambda = \sum_{\alpha > 0} k_\alpha \alpha$. Of course, $p(\lambda) = 0$ unless λ lies in the root lattice. Notice that $p = ch_{Z(0)}$

(Exercise 20.5); in particular, $p \in \mathfrak{X}$. We call p the **Kostant function** (it differs from Kostant's partition function only by a change in the sign of λ). Next, we let $q = \prod_{\alpha > 0} (\varepsilon_{\alpha/2} - \varepsilon_{-\alpha/2})$ (where the product symbol $\prod$ always denotes convolution in $\mathfrak{X}$). Call q the **Weyl function**.

We shall now prove a number of simple lemmas relating the various functions defined above. Let $f_\alpha(-k\alpha) = 1$, $f_\alpha(\lambda) = 0$ otherwise, for each positive root α, $k \in \mathbf{Z}^+$. (f_α may be thought of symbolically as $\varepsilon_0 + \varepsilon_{-\alpha} + \varepsilon_{-2\alpha} + \ldots$). It is clear that $f_\alpha \in \mathfrak{X}$.

Lemma A. (a) $p = \prod_{\alpha > 0} f_\alpha$.

        (b) $(\varepsilon_0 - \varepsilon_{-\alpha}) * f_\alpha = \varepsilon_0$.

        (c) $q = \prod_{\alpha > 0} (\varepsilon_0 - \varepsilon_{-\alpha}) * \varepsilon_\delta$.

Proof. (a) This follows at once from the definition of convolution. (b) Formally, $(\varepsilon_0 - \varepsilon_{-\alpha}) * (\varepsilon_0 + \varepsilon_{-\alpha} + \varepsilon_{-2\alpha} + \ldots) = \varepsilon_0$, since all other terms cancel. (It is easy to make this rigorous.) (c) Since $\delta = \sum \frac{1}{2}\alpha$, $\varepsilon_\delta = \prod_{\alpha > 0} \varepsilon_{\alpha/2}$. But $(\varepsilon_0 - \varepsilon_{-\alpha}) * \varepsilon_{\alpha/2} = \varepsilon_{\alpha/2} - \varepsilon_{-\alpha/2}$, so the result follows. $\square$

Lemma B. $\sigma q = (-1)^{\ell(\sigma)} q$ ($\sigma \in \mathcal{W}$), $\ell(\sigma)$ as in (10.3).

Proof. It suffices to prove this when $\sigma = \sigma_\alpha$ is a simple reflection, i.e., $\ell(\sigma) = 1$. But σ_α permutes the positive roots other than α and sends α to $-\alpha$ (Lemma 10.2B), so $\sigma_\alpha q = -q$. $\square$

Lemma C. $q * p * \varepsilon_{-\delta} = \varepsilon_0$.

Proof. Combining the three parts of Lemma A, we get $q * p * \varepsilon_{-\delta} = \prod_{\alpha > 0}$
$(\varepsilon_0 - \varepsilon_{-\alpha}) * \varepsilon_\delta * p * \varepsilon_{-\delta} = \prod_{\alpha > 0} (\varepsilon_0 - \varepsilon_\alpha) * p = \prod_{\alpha > 0} ((\varepsilon_0 - \varepsilon_\alpha) * f_\alpha) = \varepsilon_0$. $\square$

Lemma D. $ch_{Z(\lambda)}(\mu) = p(\mu - \lambda) = (p * \varepsilon_\lambda)(\mu)$.

Proof. The first equality is clear (cf. Exercise 20.5), the second is equally so. $\square$

Lemma E. $q * ch_{Z(\lambda)} = \varepsilon_{\lambda + \delta}$.

Proof. Combine Lemmas C and D. $\square$

The coefficient $(-1)^{\ell(\sigma)}$ ($\sigma \in \mathcal{W}$) which appears in Lemma B will be abbreviated henceforth to $sn(\sigma)$. Recall that when L is of type A_ℓ, $\mathcal{W}$ is isomorphic to the symmetric group $\mathscr{S}_{\ell+1}$, and $sn(\sigma)$ coincides with the sign (+ for even, − for odd) of the permutation σ.

24.2. Kostant's multiplicity formula

The idea now is to express the formal character ch_λ of the finite dimensional module $V(\lambda)$ ($\lambda \in \Lambda^+$) as $\mathbf{Z}$-linear combination of certain $ch_{Z(\mu)}$, and then use the lemmas of the preceding subsection (along with Harish-Chandra's Theorem) to simplify the result.

Let $\mathfrak{M}_\lambda$ denote the collection of all L-modules V having the following properties (for fixed $\lambda \in H^*$):

(1) V is the direct sum of weight spaces (relative to H).

(2) The action of $\mathfrak{Z}$ on V is by scalars $\chi_\lambda(z)$ ($z \in \mathfrak{Z}$), for the given $\lambda \in H^*$. ($\mathfrak{Z}$ = center of $\mathfrak{U}(L)$.)

(3) The formal character of V belongs to $\mathfrak{X}$. Of course, all standard cyclic modules of weight λ meet these criteria; so do their submodules (which are known to be not always sums of standard cyclic submodules). Indeed, $\mathfrak{M}_\lambda$ is closed under the operations of taking submodules, taking homomorphic images, and forming (finite) direct sums. In view of Harish-Chandra's Theorem (23.3), $\mathfrak{M}_\lambda = \mathfrak{M}_\mu$ precisely when λ and μ are linked.

Lemma. *Let $V \in \mathfrak{M}_\lambda$. Then V possesses at least one maximal vector (if $V \neq 0$).*

Proof. Because of property (3), for each $\alpha \succ 0$, and each weight μ of V, $\mu + k\alpha$ is not a weight of V for all sufficiently large $k \in \mathbf{Z}^+$. This makes it clear that for some weight μ, no $\mu + \alpha$ is a weight ($\alpha \succ 0$); any nonzero vector in V_μ is then maximal. $\square$

For each $\lambda \in H^*$, let $\theta(\lambda) = \{\mu \in H^* | \mu < \lambda \text{ and } \mu \sim \lambda\}$. Recall (23.2) that $\mu \sim \lambda$ means $\mu + \delta$ and $\lambda + \delta$ are $\mathscr{W}$-conjugate. In the following key result, Harish-Chandra's Theorem comes into play by limiting the possible highest weights of composition factors of $Z(\lambda)$.

Proposition. *Let $\lambda \in H^*$. Then:*

(a) $Z(\lambda)$ possesses a composition series.

(b) Each composition factor of $Z(\lambda)$ is of the form $V(\mu)$, where $\mu \in \theta(\lambda)$ and $V(\mu)$ is as defined in (20.3).

(c) $V(\lambda)$ occurs only once as a composition factor of $Z(\lambda)$.

Proof. (a) If $Z(\lambda)$ is irreducible, then $Z(\lambda) = V(\lambda)$, and there is nothing to prove. Otherwise $Z(\lambda)$ has a nonzero proper submodule V, which lies in $\mathfrak{M}_\lambda$ (the given λ being used for condition (2)). Since dim $Z(\lambda)_\lambda = 1$, λ does not occur as a weight of V. By the above lemma, V has a maximal vector (say of weight $\mu \underset{\neq}{\prec} \lambda$), hence V contains a nonzero homomorphic image W of $Z(\mu)$. In particular, $\chi_\lambda = \chi_\mu$, so $\lambda \sim \mu$ (by Harish-Chandra's Theorem), and $\mu \in \theta(\lambda)$. Consider now $Z(\lambda)/W$, W. Each module is standard cyclic (and lies in $\mathfrak{M}_\lambda$), but either has fewer weights linked to λ than $Z(\lambda)$ had, or else has the same weights linked to λ, but some of smaller multiplicity than in $Z(\lambda)$. Repetition of the above argument for $Z(\lambda)/W$ and W leads to further submodules and homomorphic images of submodules, with decreasing number of weights linked to λ or else decreasing multiplicities for those weights. This makes it evident that the process will end after finitely many steps, with a composition series for $Z(\lambda)$.

(b) Each composition factor of $Z(\lambda)$ lies in $\mathfrak{M}_\lambda$, hence possesses a maximal vector (by the lemma), hence must be standard cyclic (being irreducible). In view of (20.2), each composition factor is isomorphic to some $V(\mu)$. We saw already that μ must then belong to $\theta(\lambda)$.

(c) This is clear, since dim $Z(\lambda)_\lambda = 1$. $\square$

The proposition allows us to write $ch_{Z(\lambda)} = ch_{V(\lambda)} + \sum d(\mu)ch_{V(\mu)}$ $(d(\mu) \in \mathbf{Z}^+)$, where the summation is taken over $\mu \in \theta(\lambda)$, $\mu \neq \lambda$. With $\lambda \in H^*$ still fixed, order the elements of $\theta(\lambda)$ as $(\mu_1, \ldots, \mu_t)$, subject only to the condition that $\mu_i \prec \mu_j$ imply $i \leq j$. (In particular, $\lambda = \mu_t$.) According to the proposition, each $ch_{Z(\mu_j)}$ may in turn be expressed as a $\mathbf{Z}$-linear combination of the $ch_{V(\mu_i)}$ with $i \leq j$ ($ch_{V(\mu_j)}$ occurring with coefficient one). Therefore the resulting system of t equations has triangular matrix, relative to the chosen order, with ones on the diagonal; in particular, its determinant is 1, so we can invert it over $\mathbf{Z}$ and express each $ch_{V(\mu_j)}$ as $\mathbf{Z}$-linear combination of the $ch_{Z(\mu_i)}$ for $i \leq j$, $ch_{Z(\mu_j)}$ occurring with coefficient one. (Of course, some coefficients may now be negative.)

Corollary. Let $\lambda \in H^*$. Then $ch_{V(\lambda)}$ is a $\mathbf{Z}$-linear combination $\sum c(\mu)ch_{Z(\mu)}$ (summation over $\mu \in \theta(\lambda)$), with $c(\lambda) = 1$. $\square$

We now apply the corollary to the special case in which λ is *dominant integral*, $ch_\lambda = ch_{V(\lambda)}$. Then dim $V(\lambda)$ is finite, and $\sigma(ch_\lambda) = ch_\lambda$ for all $\sigma \in \mathscr{W}$ (Theorem 21.2). Write $ch_\lambda = \sum c(\mu)ch_{Z(\mu)}$ ($\mu \in \theta(\lambda)$) as above, with $c(\lambda) = 1$. By Lemma E of (24.1), $q * ch_\lambda = \sum c(\mu)\varepsilon_{\mu+\delta}$. By Lemma B of (24.1), $\sigma(q * ch_\lambda) = \sigma(q) * \sigma(ch_\lambda) = sn(\sigma) q * ch_\lambda$ ($\sigma \in \mathscr{W}$). On the other hand, $\sigma(\sum c(\mu)\varepsilon_{\mu+\delta}) = \sum c(\mu)\varepsilon_{\sigma^{-1}(\mu+\delta)}$. Since $\mathscr{W}$ just permutes the $\mu+\delta$ transitively (μ being linked to λ), while $c(\lambda) = 1$, it follows immediately that $c(\mu) = sn(\sigma)$ if $\sigma^{-1}(\mu+\delta) = \lambda+\delta$. Therefore

$$(*) \qquad\qquad q * ch_\lambda = \sum_{\sigma \in \mathscr{W}} sn(\sigma)\varepsilon_{\sigma(\lambda+\delta)}.$$

Finally, apply Lemma C of (24.1) to this equation to obtain: $ch_\lambda = q * p * \varepsilon_{-\delta} * ch_\lambda = p * \varepsilon_{-\delta} * (\sum_{\sigma \in \mathscr{W}} sn(\sigma) \varepsilon_{\sigma(\lambda+\delta)}) = p * (\sum_{\sigma \in \mathscr{W}} sn(\sigma) \varepsilon_{\sigma(\lambda+\delta)-\delta}) = \sum_{\sigma \in \mathscr{W}} sn(\sigma) p * \varepsilon_{\sigma(\lambda+\delta)-\delta}$.

Theorem (Kostant). Let $\lambda \in \Lambda^+$. Then the multiplicities of $V(\lambda)$ are given by the formula

$$m_\lambda(\mu) = \sum_{\sigma \in \mathscr{W}} sn(\sigma)p(\mu+\delta-\sigma(\lambda+\delta)). \quad \square$$

This formula has the virtue of expressing multiplicities directly. However, Freudenthal's recursive method (§22) is often simpler to use in practice, because summation over the Weyl group becomes very cumbersome in high rank.

24.3. Weyl's formulas

Lemma. $q = \sum_{\sigma \in \mathscr{W}} sn(\sigma)\varepsilon_{\sigma\delta}$.

Proof. This is easy to prove directly, but instead we use formula (*) of (24.2). If $\lambda = 0$, then of course $ch_\lambda = \varepsilon_0$, and the right side of (*) becomes $\sum_{\sigma \in \mathscr{W}} sn(\sigma)\varepsilon_{\sigma\delta}$. $\square$

Theorem (Weyl). *Let* $\lambda \in \Lambda^+$. *Then* $(\sum_{\sigma \in \mathscr{W}} sn(\sigma)\varepsilon_{\sigma\delta}) * ch_\lambda = \sum_{\sigma \in \mathscr{W}} sn(\sigma)\varepsilon_{\sigma(\lambda+\delta)}$.

Proof. Use formula (*) in (24.2) along with the above lemma. □

Weyl's character formula says in effect that we may calculate ch_λ as a quotient of two simple alternating sums in $\mathbf{Z}[\Lambda]$. In practice the carrying out of this "division" can be quite laborious, so Freudenthal's method (§22) is usually quicker. However, we can derive from Weyl's formula an extremely useful formula for the *dimension* of $V(\lambda)$ ($\lambda \in \Lambda^+$), which we denote by **deg (λ)**. It is clear that deg $(\lambda) = \sum_{\mu \in \Pi(\lambda)} m_\lambda(\mu)$, since $V(\lambda)$ is direct sum of weight spaces. This is just the sum of coefficients in the formal sum $\sum_\mu m_\lambda(\mu)$ $e(\mu) \in \mathbf{Z}[\Lambda]$. In the function notation, this becomes the *sum of the values* of ch_λ. Let us work now in the subalgebra $\mathfrak{X}_o \subset \mathfrak{X}$ generated by the characteristic functions ε_λ ($\lambda \in \Lambda$), so the function $v : \mathfrak{X}_o \to \mathsf{F}$ assigning to $f \in \mathfrak{X}_o$ the sum of all its values is well defined. Our problem is to compute $v(ch_\lambda)$ as a function of λ. Unfortunately, applying v to an alternating sum such as $\sum_{\sigma \in \mathscr{W}} sn(\sigma)\varepsilon_{\sigma\delta}$ gives 0, so we must proceed indirectly. Abbreviate $\sum_{\sigma \in \mathscr{W}} sn(\sigma)\varepsilon_{\sigma(\lambda+\delta)}$ by $\omega(\lambda+\delta)$, for any $\lambda \in \Lambda^+$.

The assignment $\varepsilon_\lambda \mapsto (\lambda, \alpha)\varepsilon_\lambda$ (for fixed root α) extends to an endomorphism ∂_α of $\mathfrak{X}_o$, which is in fact a derivation. The endomorphism $\partial = \prod_{\alpha \succ 0} \partial_\alpha$ is no longer a derivation, in general, but may be thought of as a differential operator. Weyl's formula reads: $\omega(\delta) * ch_\lambda = \omega(\lambda+\delta)$. Here $\omega(\delta)$ is the Weyl function (denoted q earlier), which is equal to $\varepsilon_{-\delta} * \prod_{\alpha \succ 0} (\varepsilon_\alpha - 1)$ (cf. the original definition of q). Now apply ∂ to both sides of Weyl's formula, using the Leibniz rule for derivations ∂_α, to obtain: $v(\partial(\omega(\delta) * ch_\lambda)) = v(\partial\omega(\delta))$ $v(ch_\lambda)$, i.e., deg $(\lambda) = \dfrac{v(\partial(\omega(\lambda+\delta)))}{v(\partial\omega(\delta))}$.

A moment's thought shows that $v(\partial\varepsilon_\delta) = \prod_{\alpha \succ 0} (\delta, \alpha)$; similarly, $v(\partial\varepsilon_{\sigma\delta}) = \prod_{\alpha \succ 0}$ $(\sigma\delta, \alpha) = \prod_{\alpha \succ 0} (\delta, \sigma^{-1}\alpha)$. But recall that the number of positive roots sent to negative roots by σ^{-1} is $\ell(\sigma^{-1}) = \ell(\sigma)$ (cf. Lemma A (10.3)), so this is just $sn(\sigma) \prod_{\alpha \succ 0} (\delta, \alpha)$, $sn(\sigma) = (-1)^{\ell(\sigma)}$. In other words, $v(\partial\omega(\delta)) = \sum_{\sigma \in \mathscr{W}} sn(\sigma) \, v(\partial\varepsilon_{\sigma\delta})$ $= \sum_{\sigma \in \mathscr{W}} sn(\sigma)^2 \prod_{\alpha \succ 0} (\delta, \alpha) = \text{Card} (\mathscr{W}) \prod_{\alpha \succ 0} (\delta, \alpha)$. The same argument for $\omega(\lambda+\delta)$ leads to Card $(\mathscr{W}) \prod_{\alpha \succ 0} (\lambda+\delta, \alpha)$. Forming the quotient we therefore have:

Corollary. *Let* $\lambda \in \Lambda^+$. *Then deg* $(\lambda) = \dfrac{\prod_{\alpha \succ 0} (\lambda+\delta, \alpha)}{\prod_{\alpha \succ 0} (\delta, \alpha)}$. □

In order to compute some examples, we observe that after multiplying both numerator and denominator by $\prod_{\alpha \succ 0} \dfrac{2}{(\alpha, \alpha)}$, we get deg $(\lambda) = \dfrac{\prod_{\alpha \succ 0} \langle\lambda+\delta, \alpha\rangle}{\prod_{\alpha \succ 0} \langle\delta, \alpha\rangle}$ (a quotient of integers). But $<\lambda+\delta, \alpha> = (\lambda+\delta, \alpha^\vee)$, $\alpha^\vee$ the *dual* root. In turn, since $\Delta^{\vee}$ is a base of $\Phi^{\vee}$ (Exercise 10.1), we can write $\alpha^{\vee} = \sum_i c_i^{(\alpha)} \alpha_i^{\vee}$, whence $<\lambda + \delta, \alpha> = \sum_i c_i^{(\alpha)} <\lambda + \delta, \alpha_i> = \sum_i c_i^{(\alpha)}(m_i + 1)$ $(\lambda = \sum_i m_i \lambda_i)$. So we need only compute the integers $c_i^{(\alpha)}$ (Exercise 7).

Examples. For type A_1, $\lambda_1 = \frac{1}{2}\alpha = \delta$, so the formula reads: deg $(\lambda) = m+1$, $\lambda = m\lambda_1$, cf. Theorem 7.2.

Let us concentrate now on rank 2, writing $\lambda = m_1\lambda_1 + m_2\lambda_2$. For type A_2 the positive roots are α_1, α_2, $\alpha_1 + \alpha_2$. Accordingly, the denominator above equals $1 \cdot 1 \cdot 2$, while the numerator is $(m_1 + 1)(m_2 + 1)(m_1 + m_2 + 2)$. For B_2 and G_2 the calculations are similar (take α_2 short for B_2, α_1 short for G_2, in conformity with §11). To summarize (Exercise 7):

(A_2) $\dfrac{1}{2}(m_1 + 1)(m_2 + 1)(m_1 + m_2 + 2)$

(B_2) $\dfrac{1}{3!}(m_1 + 1)(m_2 + 1)(m_1 + m_2 + 2)(2m_1 + m_2 + 3)$

(G_2) $\dfrac{1}{5!}(m_1 + 1)(m_2 + 1)(m_1 + m_2 + 2)(m_1 + 2m_2 + 3)(m_1 + 3m_2 + 4)(2m_1 + 3m_2 + 5)$.

For G_2, deg $(\lambda_2) = 14$. Since $\lambda_2 = 3\alpha_1 + 2\alpha_2$, the highest root, we recognize $V(\lambda_2)$ as the module for the adjoint representation. Deg $(\lambda_1) = 7$. Here $V(\lambda_1)$ is $\mathfrak{C}_o$ (the trace 0 subspace of the Cayley algebra (19.3)).

24.4. Steinberg's formula

We can combine the formulas of Kostant and Weyl to obtain an explicit formula (due to R. Steinberg) for the number of occurrences of $V(\lambda)$ in the tensor product $V(\lambda') \otimes V(\lambda'')$. Because of Weyl's Theorem on complete reducibility of finite dimensional L-modules (6.3), if λ', $\lambda'' \in \Lambda^+$, we can write $V(\lambda') \otimes V(\lambda'')$ as direct sum of certain $V(\lambda)$, each occurring $n(\lambda)$ times (write $n(\lambda) = 0$ if $V(\lambda)$ does not occur at all). In particular, the formal character of the tensor product equals $\sum_{\lambda \in \Lambda^+} n(\lambda) \, ch_\lambda$. On the other hand, we proved in (22.5) that this formal character equals $ch_{\lambda'} * ch_{\lambda''}$. Thus:

(1) $$ch_{\lambda'} * ch_{\lambda''} = \sum_\lambda n(\lambda) ch_\lambda.$$

As before, let us abbreviate the expression $\sum_{\sigma \in \mathscr{W}} sn(\sigma)\varepsilon_{\sigma(\mu+\delta)}$ to $\omega(\mu+\delta)$, for $\mu \in \Lambda^+$. If we multiply both sides of (1) by $\omega(\delta)$ and use Weyl's formula (24.3) for λ'', λ (respectively), we get:

(2) $$ch_{\lambda'} * \omega(\lambda'' + \delta) = \sum_\lambda n(\lambda)\omega(\lambda + \delta).$$

Next we write $ch_{\lambda'} = \sum_\mu m_{\lambda'}(\mu)\varepsilon_\mu$ and replace $m_{\lambda'}(\mu)$ by its value in Kostant's formula (24.2). Equation (2) then becomes:

(3) $$\sum_\mu \sum_{\sigma \in \mathscr{W}} sn(\sigma)p(\mu + \delta - \sigma(\lambda' + \delta))\varepsilon_\mu * \omega(\lambda'' + \delta) = \sum_\lambda n(\lambda)\omega(\lambda + \delta).$$

Using the explicit form of $\omega(\lambda'' + \delta)$, this reads:

$$(4) \quad \sum_{\mu} \sum_{\sigma \in \mathscr{W}} \sum_{\tau \in \mathscr{W}} sn(\sigma\tau)p(\mu + \delta - \sigma(\lambda' + \delta))\varepsilon_{\tau(\mu + \delta) + \mu}$$

$$= \sum_{\lambda} \sum_{\sigma \in \mathscr{W}} n(\lambda)sn(\sigma)\varepsilon_{\sigma(\lambda + \delta)}.$$

To compare the two sides of (4) we first change variables. On the right, replace λ by ν, where $\sigma(\lambda + \delta) = \nu + \delta$, to get

$$(5) \quad \sum_{\nu} \sum_{\sigma \in \mathscr{W}} sn(\sigma)n(\sigma(\nu + \delta) - \delta)\varepsilon_{\nu + \delta}.$$

On the left, replace μ by ν, where $\tau(\lambda'' + \delta) + \mu = \nu + \delta$, obtaining:

$$(6) \quad \sum_{\nu} \sum_{\sigma \in \mathscr{W}} \sum_{\tau \in \mathscr{W}} sn(\sigma\tau)p(\nu + 2\delta - \sigma(\lambda' + \delta) - \tau(\lambda'' + \delta))\varepsilon_{\nu + \delta}.$$

Now let ν be *dominant*. Then $\sigma(\nu + \delta) - \delta$ cannot be dominant unless $\sigma = 1$ (Exercise 13.10). Therefore, $n(\sigma(\nu + \delta) - \delta) = 0$ unless $\sigma = 1$, which means that the coefficient of $\varepsilon_{\nu + \delta}$ in (5) is precisely $n(\nu)$. In view of (6), we have proved:

Theorem (Steinberg). *Let* λ', $\lambda'' \in \Lambda^+$. *Then the number of times* $V(\lambda)$, $\lambda \in \Lambda^+$, *occurs in* $V(\lambda') \otimes V(\lambda'')$ *is given by the formula*

$$\sum_{\sigma \in \mathscr{W}} \sum_{\tau \in \mathscr{W}} sn(\sigma\tau)p(\lambda + 2\delta - \sigma(\lambda' + \delta) - \tau(\lambda'' + \delta)). \quad \square$$

This formula (like Kostant's) is very explicit, but not at all easy to apply when the Weyl group is large. A formula which is often more practical is developed in Exercise 9.

Exercises

1. Give a direct proof of Weyl's character formula (24.3) for type A_1.
2. Use Weyl's dimension formula to show that a faithful irreducible finite dimensional L-module of smallest possible dimension has highest weight λ_i for some $1 \leq i \leq \ell$.
3. Use Kostant's formula to check some of the multiplicities listed in Example 1 (22.4), and compare ch_λ there with the expression given by Weyl's formula.
4. Compare Steinberg's formula for the special case A_1 with the Clebsch-Gordan formula (Exercise 22.7).
5. Using Steinberg's formula, decompose the G_2-module $V(\lambda_1) \otimes V(\lambda_2)$ into its irreducible constituents. Check that the dimensions add up correctly to the product $\dim V(\lambda_1) \cdot \dim V(\lambda_2)$, using Weyl's formula.
6. Let $L = \mathfrak{sl}(3, F)$. Abbreviate $\lambda = m_1\lambda_1 + m_2\lambda_2$ by (m_1, m_2). Use Steinberg's formula to verify that $V(1, 0) \otimes V(0, 1) \cong V(0, 0) \oplus V(1, 1)$.
7. Verify the degree formulas in (24.3); derive such a formula for type C_3. How can the integers $c_i^{(\alpha)}$ be found in general?

8. Let $\lambda \in \Lambda$. If there exists $\sigma \neq 1$ in $\mathscr{W}$ fixing λ, prove that $\sum\limits_{\sigma(\lambda)=\lambda} sn(\sigma)\varepsilon_{\sigma(\lambda)}$
$= 0$. [Use the fact that λ lies in the closure but not the interior of some Weyl chamber to find a reflection fixing λ, and deduce that the group fixing λ has even order.]

9. The purpose of this exercise is to obtain another decomposition of a tensor product, based on explicit knowledge of the weights of one module involved. Begin, as in (24.4), with the equation (2) $ch_{\lambda'} * \omega(\lambda'' + \delta)$
$= \sum\limits_{\lambda \in \Lambda^+} n(\lambda)\omega(\lambda + \delta)$. Replace $ch_{\lambda'}$ on the left side by $\sum\limits_{\lambda \in \Lambda} m_{\lambda'}(\lambda)\varepsilon_\lambda$, and combine
to get: $\sum\limits_{\sigma \in \mathscr{W}} sn(\sigma) \sum\limits_\lambda m_{\lambda'}(\lambda)\varepsilon_{\sigma(\lambda + \lambda'' + \delta)}$, using the fact that $\mathscr{W}$ permutes weight spaces of $V(\lambda')$. Next show that the right side of (2) can be expressed as $\sum\limits_{\sigma \in \mathscr{W}} sn(\sigma) \sum\limits_{\lambda \in \Lambda^+} n(\lambda)\varepsilon_{\sigma(\lambda + \delta)}$. Define $t(\mu)$ to be 0 if some element $\sigma \neq 1$ of $\mathscr{W}$ fixes μ, and to be $sn(\sigma)$ if nothing but 1 fixes μ and if $\sigma(\mu)$ is dominant. Then deduce from Exercise 8 that:

$$ch_{\lambda'} * ch_{\lambda''} = \sum\limits_{\lambda \in \Pi(\lambda')} m_{\lambda'}(\lambda)t(\lambda + \lambda'' + \delta)ch_{\{\lambda + \lambda'' + \delta\} - \delta},$$

where the braces denote the unique dominant weight to which the indicated weight is conjugate.

10. Rework Exercises 5, 6, using the approach of Exercise 9.

11. With notation as in Exercise 6, verify that $V(1, 1) \otimes V(1, 2) \cong V(2, 3)$
$\oplus V(3, 1) \oplus V(0, 4) \oplus V(1, 2) \oplus V(1, 2) \oplus V(2, 0) \oplus V(0, 1)$.

12. Deduce from Steinberg's formula that the only possible $\lambda \in \Lambda^+$ for which $V(\lambda)$ can occur as a summand of $V(\lambda') \otimes V(\lambda'')$ are those of the form $\mu + \lambda''$, where $\mu \in \Pi(\lambda')$. In case all such $\mu + \lambda''$ are *dominant*, deduce from Exercise 9 that $V(\mu + \lambda'')$ does occur in the tensor product, with multiplicity $m_{\lambda'}(\mu)$. Using these facts, decompose $V(1, 3) \otimes V(4, 4)$ for type A_2 (cf. Example 1 of (22.4)).

Notes

Weyl's original proofs used integration on compact Lie groups; later Freudenthal devised a more algebraic (but less intuitive) proof: see Freudenthal–deVries [1], Jacobson [1], Samelson [1]. The present approach is suggested by work of Verma [1], and follows closely a recent paper by Bernstein, Gel'fand, Gel'fand [1]. See Kostant [1] for the original (rather complicated) proof of his formula, and Cartier [1] for simplifying remarks. Steinberg's formula is derived concisely in Steinberg [1]. The approach to tensor products sketched in Exercise 9 is due to Klimyk [1], based on older work of Brauer and Weyl. Exercise 12 was pointed out to the author by D.-N. Verma.

Chapter VII

Chevalley Algebras and Groups

The notation is that of earlier chapters. L is a semisimple Lie algebra over the algebraically closed field F of characteristic 0, H a CSA, Φ the root system.

In this chapter we shall see how to construct L and its irreducible representations "over $\mathbf{Z}$", a possibility which has been more or less apparent all along in case L is classical. The results to be obtained actually go much further, enabling us to construct "Chevalley groups" and representations of these groups over arbitrary fields. This is a large subject, to which we can only introduce the reader.

25. Chevalley basis of L

25.1. Pairs of roots

It will be proved in (25.2) that L has a basis for which the structure constants are integral. But first we must establish some facts about pairs of roots α, β for which $\alpha + \beta$ is also a root, with an eye toward the equation $[x_\alpha x_\beta] = c_{\alpha\beta} x_{\alpha+\beta}$. The following proposition depends only on the root system Φ (not on L).

Proposition. *Let α, β be linearly independent roots, $\beta - r\alpha, \ldots, \beta, \ldots, \beta + q\alpha$ the α-string through β. Then:*

(a) $\langle \beta, \alpha \rangle = r - q$.

(b) *At most two root lengths occur in this string.*

(c) *If $\alpha + \beta \in \Phi$, then $r + 1 = \dfrac{q(\alpha \mid \beta, \alpha \mid \beta)}{(\beta, \beta)}$.*

Proof. (a) This was proved in (9.4) (as well as in Proposition 8.4(e), via the representation theory of $\mathfrak{sl}(2, \mathsf{F})$).

(b) $\Phi' = (\mathbf{Z}\alpha + \mathbf{Z}\beta) \cap \Phi$ is a root system of rank 2 (in the subspace of E spanned by α, β): cf. Exercise 9.7. If reducible, it must be of type $\mathsf{A}_1 \times \mathsf{A}_1$, i.e., $\Phi' = \{\pm\alpha, \pm\beta\}$, and there is nothing to prove. If irreducible, $\Phi' = \mathsf{A}_2$, B_2, or G_2, and the result follows (alternatively, use Lemma 10.4C).

(c) This can be done by inspecting the root systems of rank 2 (Exercise 1). However, the following geometric argument works in general. First, we deduce from (a):

$$(r+1) - \frac{q(\alpha+\beta, \alpha+\beta)}{(\beta, \beta)} = q + \frac{2(\beta, \alpha)}{(\alpha, \alpha)} + 1 - \frac{q(\alpha+\beta, \alpha+\beta)}{(\beta, \beta)}$$

$$= \frac{2(\beta, \alpha)}{(\alpha, \alpha)} + 1 - \frac{q(\alpha, \alpha)}{(\beta, \beta)} - \frac{2q(\alpha, \beta)}{(\beta, \beta)}$$

$$= (\langle \beta, \alpha \rangle + 1)\left(1 - \frac{q(\alpha, \alpha)}{(\beta, \beta)}\right).$$

Call the respective factors of this last product A, B. We have to show that one or the other is 0. The situation here is *not* symmetric in α, β, so two cases must be distinguished:

Case i: $(\alpha, \alpha) \geq (\beta, \beta)$. Then $|\langle \beta, \alpha \rangle| \leq |\langle \alpha, \beta \rangle|$. Since α, β are independent, we know (9.4) that $\langle \beta, \alpha \rangle \langle \alpha, \beta \rangle = 0$, 1, 2, or 3. The inequality forces $\langle \beta, \alpha \rangle = -1$, 0, or 1. In the first case, $A = 0$ and we're done. Otherwise $(\beta, \alpha) \geq 0$, so $(\beta+\alpha, \beta+\alpha)$ is strictly larger than both (β, β) and (α, α). Since $\alpha+\beta \in \Phi$, (b) implies that $(\alpha, \alpha) = (\beta, \beta)$. Similarly, $(\beta+2\alpha, \beta+2\alpha) > (\beta+\alpha, \beta+\alpha)$, so (b) implies that $\beta+2\alpha \notin \Phi$, i.e., that $q = 1$, forcing $B = 0$.

Case ii: $(\alpha, \alpha) < (\beta, \beta)$. Then $(\alpha+\beta, \alpha+\beta) = (\alpha, \alpha)$ or (β, β) (by (b)), forcing in either case $(\alpha, \beta) < 0$ (hence $\langle \alpha, \beta \rangle < 0$). In turn, $(\beta-\alpha, \beta-\alpha) > (\beta, \beta) > (\alpha, \alpha)$, so $\beta-\alpha \notin \Phi$ (by (b) again), i.e., $r = 0$. As before, $\langle \alpha, \beta \rangle \langle \beta, \alpha \rangle = 0$, 1, 2, or 3, but here we have $|\langle \alpha, \beta \rangle| < |\langle \beta, \alpha \rangle|$, forcing $\langle \alpha, \beta \rangle = -1$, 0, or 1. But we know that $\langle \alpha, \beta \rangle < 0$, so $\langle \alpha, \beta \rangle = -1$. By (a),
$q = -\langle \beta, \alpha \rangle = \frac{\langle \beta, \alpha \rangle}{\langle \alpha, \beta \rangle} = \frac{(\beta, \beta)}{(\alpha, \alpha)}$, whence $B = 0$.   ☐

25.2. Existence of a Chevalley basis

Lemma. *Let α, β be independent roots. Choose $x_\alpha \in L_\alpha$, $x_{-\alpha} \in L_{-\alpha}$ for which $[x_\alpha x_{-\alpha}] = h_\alpha$, and let $x_\beta \in L_\beta$ be arbitrary. Then if $\beta - r\alpha, \ldots, \beta + q\alpha$ is the α-string through β, we have: $[x_{-\alpha}[x_\alpha x_\beta]] = q(r+1)x_\beta$. (For the definition of h_α, see Proposition 8.3.)*

Proof. If $\alpha+\beta \notin \Phi$, then $q = 0$ and $[x_\alpha x_\beta] = 0$, so both sides of the above equation are 0. In general, we can exploit the adjoint representation of S_α ($\cong \mathfrak{sl}(2, F)$) on L, as we did for an arbitrary representation in (22.2). Namely, the S_α-submodule of L generated by x_β has dimension $r+q+1$ (the number of roots in the α-string through β), highest weight $r+q$. In the notation of Lemma 7.2, x_β is a (nonzero) multiple of v_q, and the successive application of ad x_α, ad $x_{-\alpha}$ multiplies v_q (hence also x_β) by the scalar $q(r+1)$.   ☐

Proposition. *It is possible to choose root vectors $x_\alpha \in L_\alpha$ ($\alpha \in \Phi$) satisfying:*

(a) $[x_\alpha x_{-\alpha}] = h_\alpha$.
(b) If α, β, $\alpha+\beta \in \Phi$, $[x_\alpha x_\beta] = c_{\alpha\beta}x_{\alpha+\beta}$, then $c_{\alpha\beta} = -c_{-\alpha, -\beta}$. For any such choice of root vectors, the scalars $c_{\alpha\beta}(\alpha, \beta, \alpha+\beta \in \Phi)$ automatically satisfy:

(c) $c_{\alpha\beta}^2 = q(r+1) \dfrac{(\alpha+\beta, \alpha+\beta)}{(\beta, \beta)}$, where $\beta-r\alpha, \ldots, \beta+q\alpha$ is the α-string through β.

Proof. Recall (Proposition 14.3) that L possesses an automorphism σ of order 2 sending L_α to $L_{-\alpha}$ ($\alpha \in \Phi$) and acting on H as multiplication by -1. For arbitrary nonzero $x_\alpha \in L_\alpha$, $x_{-\alpha} = -\sigma(x_\alpha) \in L_{-\alpha}$ is nonzero, and $\kappa(x_\alpha, x_{-\alpha}) \neq 0$ (κ the Killing form). Replacing x_α by cx_α ($c \in \mathsf{F}$) multiplies this value by c^2. Since F is algebraically closed, it is therefore possible to modify the choice so that $\kappa(x_\alpha, x_{-\alpha})$ takes any prescribed nonzero value. We specify $\kappa(x_\alpha, x_{-\alpha}) = \dfrac{2}{(\alpha, \alpha)}$. According to Proposition 8.3(c), this forces $[x_\alpha x_{-\alpha}] = h_\alpha \left(= \dfrac{2t_\alpha}{(\alpha, \alpha)} \right)$. For each pair of roots $\{\alpha, -\alpha\}$ we fix such a choice of $\{x_\alpha, x_{-\alpha}\}$, so (a) is satisfied.

Now let $\alpha, \beta, \alpha+\beta \in \Phi$, so $[x_\alpha x_\beta] = c_{\alpha\beta} x_{\alpha+\beta}$ for some $c_{\alpha\beta} \in \mathsf{F}$. Applying σ to this equation, we get $[-x_{-\alpha}, -x_{-\beta}] = -c_{\alpha\beta} x_{-\alpha-\beta}$. On the other hand, $[x_{-\alpha} x_{-\beta}] = c_{-\alpha, -\beta} x_{-\alpha-\beta}$, so (b) follows.

Having chosen root vectors $\{x_\alpha, \alpha \in \Phi\}$ satisfying (a) (b), consider the situation: $\alpha, \beta, \alpha+\beta \in \Phi$ (in particular, α and β, hence t_α and t_β (cf. (8.2)), are linearly independent). Since $t_{\alpha+\beta} = t_\alpha + t_\beta$, it follows from (a) that $[c_{\alpha\beta} x_{\alpha+\beta}, c_{\alpha\beta} x_{-\alpha-\beta}] = c_{\alpha\beta}^2 h_{\alpha+\beta} = \dfrac{2c_{\alpha\beta}^2}{(\alpha+\beta, \alpha+\beta)}(t_\alpha + t_\beta)$. On the other hand, (b) implies that the left side also equals $-[[x_\alpha x_\beta]\ [x_{-\alpha} x_{-\beta}]] = -[x_\alpha[x_\beta[x_{-\alpha} x_{-\beta}]]] + [x_\beta[x_\alpha[x_{-\alpha} x_{-\beta}]]]] = [x_\alpha[x_\beta[x_{-\beta} x_{-\alpha}]]] + [x_\beta[x_\alpha[x_{-\alpha} x_{-\beta}]]]]$. Let the β-string through α be $\alpha-r'\beta, \ldots, \alpha+q'\beta$. Then the above lemma may be applied to each term (after replacing α, β by their negatives, which does not affect r, q, r', q') to obtain: $q'(r'+1) [x_\alpha x_{-\alpha}] + q(r+1) [x_\beta x_{-\beta}] = \dfrac{2q'(r'+1)}{(\alpha, \alpha)} t_\alpha + \dfrac{2q(r+1)}{(\beta, \beta)} t_\beta$. Comparing these coefficients with those above, and using the linear independence of t_α and t_β, we get (c). $\square$

We are now in a position to construct a **Chevalley basis** of L. This is by definition any basis $\{x_\alpha, \alpha \in \Phi; h_i, 1 \leq i \leq \ell\}$ for which the x_α satisfy (a) (b) of the preceding proposition, while $h_i = h_{\alpha_i}$ for some base $\Delta = \{\alpha_1, \ldots, \alpha_\ell\}$ of Φ.

Theorem (Chevalley). *Let $\{x_\alpha, \alpha \in \Phi; h_i, 1 \leq i \leq \ell\}$ be a Chevalley basis of L. Then the resulting structure constants lie in $\mathbf{Z}$. More precisely:*

(a) $[h_i h_j] = 0, 1 \leq i, j \leq \ell$.

(b) $[h_i x_\alpha] = \langle \alpha, \alpha_i \rangle x_\alpha, 1 \leq i \leq \ell, \alpha \in \Phi$.

(c) $[x_\alpha x_{-\alpha}] = h_\alpha$ *is a $\mathbf{Z}$-linear combination of $h_1, \ldots, h_\ell$.*

(d) *If α, β are independent roots, $\beta-r\alpha, \ldots, \beta+q\alpha$ the α-string through β, then $[x_\alpha x_\beta] = 0$ if $q = 0$, while $[x_\alpha x_\beta] = \pm(r+1)x_{\alpha+\beta}$ if $\alpha+\beta \in \Phi$.*

Proof. (a) is clear, while (b) follows from the fact that $\alpha(h_i) = \langle \alpha, \alpha_i \rangle$.

As to (c), recall that the dual roots $\alpha^\vee = \dfrac{2\alpha}{(\alpha, \alpha)}$ form a root system, with base $\Delta^\vee = \{\alpha_1^\vee, \ldots, \alpha_\ell^\vee\}$ (Exercise 10.1). Under the Killing form identification of H with H^*, t_α corresponds to α and h_α to $\alpha^\vee$. Since each $\alpha^\vee$ is a $\mathbf{Z}$-linear combination of $\Delta^\vee$, each h_α is a $\mathbf{Z}$-linear combination of $h_1, \ldots, h_\ell$. Finally, (d) follows from part (c) of the preceding proposition, combined with part (c) of Proposition 25.1.   □

It may seem strange to the reader that we have required $c_{\alpha\beta} = -c_{-\alpha, -\beta}$ rather than $c_{\alpha\beta} = c_{-\alpha, -\beta}$ in our definition of Chevalley basis. However, this asymmetry is inevitable: Given condition (a) of the proposition, it can be shown by skillful use of the Jacobi identity that $c_{\alpha\beta} c_{-\alpha, -\beta} = -(r+1)^2$, which implies that condition (d) of the theorem could not hold unless we had condition (b) of the proposition. (This was Chevalley's original line of argument.) The reader should verify (Exercise 2) that the bases given in (1.2) for the classical algebras can be modified to yield Chevalley bases. Chevalley's theorem has the virtue of providing a uniform existence proof for Chevalley bases, as well as specifying how the structure constants arise from the root system.

25.3. Uniqueness questions

How unique is a Chevalley basis? Once Δ is fixed, the h_i are completely determined. On the other hand, it is possible to vary somewhat the choice of the x_α. Say x_α is replaced by $\eta(\alpha)x_\alpha$ ($\alpha \in \Phi$). Then $[\eta(\alpha)x_\alpha, \eta(-\alpha)x_{-\alpha}] = \eta(\alpha)\eta(-\alpha)h_\alpha$, so we must have (*) $\eta(\alpha)\eta(-\alpha) = 1$ in order to satisfy condition (a) of Proposition 25.2. If α, β, $\alpha+\beta \in \Phi$, then $[\eta(\alpha)x_\alpha, \eta(\beta)x_\beta] = \eta(\alpha)\eta(\beta) [x_\alpha x_\beta] = c_{\alpha\beta}\eta(\alpha)\eta(\beta)x_{\alpha+\beta} = c'_{\alpha\beta}\eta(\alpha+\beta)x_{\alpha+\beta}$, where $c'_{\alpha\beta} = \dfrac{c_{\alpha\beta}\eta(\alpha)\eta(\beta)}{\eta(\alpha+\beta)}$.
To satisfy (b) of Proposition 25.2, a similar calculation, using (*), shows that we must also have $c'_{\alpha\beta} = \dfrac{c_{\alpha\beta}\eta(\alpha+\beta)}{\eta(\alpha)\eta(\beta)}$, or in other words, (**) $\eta(\alpha)\eta(\beta) = \pm\eta(\alpha+\beta)$. Conversely, it is clear that any function $\eta: \Phi \to \mathsf{F}$ satisfying (*) and (**) can be used to modify the choice of the x_α.

The question of *signs* is more delicate. We have $[x_\alpha x_\beta] = \pm(r+1)x_{\alpha+\beta}$ (α, β, $\alpha+\beta \in \Phi$), but the argument used to establish this equation left unsettled the choice of plus or minus. This is not accidental, as the reader can see by choosing $\begin{pmatrix} 0 & 0 & -1 \\ 0 & 0 & 0 \\ 0 & 0 & 0 \end{pmatrix}$ in place of $\begin{pmatrix} 0 & 0 & 1 \\ 0 & 0 & 0 \\ 0 & 0 & 0 \end{pmatrix}$ as part of a Chevalley basis for $\mathfrak{sl}(3, \mathsf{F})$: there is no reason to prefer one choice over the other. There does exist an algorithm for making a consistent choice of signs, based on knowledge of Φ alone, and this leads to yet another proof of the *isomorphism theorem* (14.2, 18.4) (see Notes for this section). Of course, such a proof is circular unless the existence of the automorphism σ is established independently!

We remark that one can also prove the *existence* of L (18.4) by constructing the multiplication table explicitly and then verifying the Jacobi identity. Such a proof has been written down by Tits (see Notes below); though "elementary" in character, it is quite lengthy compared to the proof based on Serre's Theorem (18.4).

25.4. Reduction modulo a prime

The $\mathbf{Z}$-span $L(\mathbf{Z})$ of a Chevalley basis $\{x_\alpha, h_i\}$ is a lattice in L, independent of the choice of Δ. It is even a Lie algebra over $\mathbf{Z}$ (in the obvious sense) under the bracket operation inherited from L (closure being guaranteed by Theorem 25.2). If $\mathbf{F}_p = \mathbf{Z}/p\mathbf{Z}$ is the prime field of characteristic p, then the tensor product $L(\mathbf{F}_p) = L(\mathbf{Z}) \otimes_\mathbf{Z} \mathbf{F}_p$ is defined: $L(\mathbf{F}_p)$ is a vector space over $\mathbf{F}_p$ with basis $\{x_\alpha \otimes 1, h_i \otimes 1\}$. Moreover, the bracket operation in $L(\mathbf{Z})$ induces a natural Lie algebra structure on $L(\mathbf{F}_p)$. The multiplication table is essentially the same as the one given in Theorem 25.2, with integers reduced mod p.

If K is any field extension of $\mathbf{F}_p$ then $L(\mathsf{K}) = L(\mathbf{F}_p) \otimes_{\mathbf{F}_p} \mathsf{K} = L(\mathbf{Z}) \otimes_\mathbf{Z} \mathsf{K}$ inherits both basis and Lie algebra structure from $L(\mathbf{F}_p)$. In this way we associate with the pair (L, K) a Lie algebra over K whose structure resembles that of L. We call $L(\mathsf{K})$ a **Chevalley algebra**. Even though $L(\mathbf{Z})$ depends on how the root vectors x_α are chosen, it is easily seen (Exercise 5) to be defined up to isomorphism (over $\mathbf{Z}$) by L alone; similarly, the algebra $L(\mathsf{K})$ depends (up to isomorphism) only on the pair (L, K).

To illustrate these remarks, we consider $L = \mathfrak{sl}(\ell+1, \mathsf{F})$. It is clear that $L(\mathsf{K})$ has precisely the same multiplication table as $\mathfrak{sl}(\ell+1, \mathsf{K})$, relative to the standard basis (1.2). So $L(\mathsf{K}) \cong \mathfrak{sl}(\ell+1, \mathsf{K})$. The only real change that takes place in passing from F to K is that $L(\mathsf{K})$ *may fail to be simple*: here it has one dimensional center consisting of scalar matrices whenever char K divides $\ell+1$ (cf. Exercise 2.3 and Exercise 8 below).

25.5. Construction of Chevalley groups (adjoint type)

Proposition. *Let* $\alpha \in \Phi$, $m \in \mathbf{Z}^+$. *Then* $(\operatorname{ad} x_\alpha)^m/m!$ *leaves* $L(\mathbf{Z})$ *invariant.*

Proof. It suffices to show that each element of the Chevalley basis is sent back into $L(\mathbf{Z})$. We have $(\operatorname{ad} x_\alpha)(h_i) = [x_\alpha h_i] = -\langle \alpha, \alpha_i \rangle x_\alpha \in L(\mathbf{Z})$, while $(\operatorname{ad} x_\alpha)^m/m! \ (h_i) - 0$ for all $m \geq 2$. Similarly, $(\operatorname{ad} x_\alpha)(x_{-\alpha}) = h_\alpha \in L(\mathbf{Z})$. $(\operatorname{ad} x_\alpha)^2/2 \cdot (x_{-\alpha}) = \frac{1}{2}[x_\alpha h_\alpha] = -x_\alpha \in L(\mathbf{Z})$ and $\dfrac{(\operatorname{ad} x_\alpha)^m}{m!} (x_{-\alpha}) = 0$ for all $m \geq 3$. Of course, $\dfrac{(\operatorname{ad} x_\alpha)^m}{m!} (x_\alpha) = 0$ for $m \geq 1$. It remains to consider the basis elements x_β, $\beta \neq \pm\alpha$. If $\beta - r\alpha, \ldots, \beta + q\alpha$ is the α-string through β, then the integers which play the role of r for the roots $\beta+\alpha, \beta+2\alpha, \ldots, \beta+q\alpha$ are (respectively) $r+1, r+2, \ldots, r+q$. Therefore,

$$\frac{(\operatorname{ad} x_\alpha)^m}{m!} (x_\beta) = \pm \frac{(r+1)(r+2)\ldots(r+m)}{m!} x_{\beta+m\alpha} \text{ (or 0, if } \beta+m\alpha \notin \Phi).$$

The coefficient just involves the binomial coefficient $\binom{r+m}{m}$, so the right side is an integral multiple of $x_{\beta+m\alpha}$. $\square$

The proposition has the following significance. L is a module for the adjoint representation of L, and $L(\mathbf{Z})$ is a lattice in L stable under the endomorphisms $(\text{ad } x_\alpha)^m/m!$, hence also under $\exp \text{ad } x_\alpha = 1 + \text{ad } x_\alpha + (\text{ad } x_\alpha)^2/2! + \ldots$ (the sum being finite, because $\text{ad } x_\alpha$ is nilpotent). Relative to the Chevalley basis, $\text{Int } L = G$ can be thought of as a group of matrices. The subgroup generated by all $\exp \text{ad } cx_\alpha$ ($\alpha \in \Phi$, $c \in \mathbf{Z}$) leaves $L(\mathbf{Z})$ invariant, hence consists of matrices with integral coefficients (and determinant 1). In particular, if p is a prime, $\mathbf{F}_p$ the prime field of p elements, then reducing all matrix entries mod p yields a matrix group over $\mathbf{F}_p$ which acts on the Lie algebra $L(\mathbf{F}_p)$ as a group of automorphisms, denoted $G \mathbf{F}_p)$.

More generally, let T be an indeterminate. The matrix group generated by all $\exp \text{ad } Tx_\alpha$ ($\alpha \in \Phi$) consists of matrices with coefficients in $\mathbf{Z}[T]$ (and determinant 1), so that specializing T to elements of an arbitrary extension field K of $\mathbf{F}_p$ yields a matrix group $G(\text{K})$ over K. Such a group is called a **Chevalley group (of adjoint type)**. When K is finite, the group is finite and (apart from a few exceptions) *simple*; by proving this, Chevalley was able to exhibit several families of previously unknown finite simple groups.

Exercises

1. Prove Proposition 25.1(c) by inspecting root systems of rank 2. [Note that one of α, β may be assumed simple.]
2. How can the bases for the classical algebras exhibited in (1.2) be modified so as to obtain Chevalley bases? [Cf. Exercise 14.7.]
3. Use the proof of Proposition 25.2 to give a new proof of Exercise 9.10.
4. If only one root length occurs in each component of Φ (i.e., Φ has irreducible components of types A, D, E), prove that all $c_{\alpha\beta} = \pm 1$ in Theorem 25.2 (when α, β, $\alpha+\beta \in \Phi$).
5. Prove that different choices of Chevalley basis for L lead to isomorphic Lie algebras $L(\mathbf{Z})$ over $\mathbf{Z}$. ("Isomorphism over $\mathbf{Z}$" is defined just as for a field.)
6. For the algebra of type B_2, let the positive roots be denoted α, β, $\alpha+\beta$, $2\beta+\alpha$. Check that the following equations are those resulting from a Chevalley basis (in particular, the signs $\pm$ are consistent):

$$[h_\beta, x_\beta] = 2x_\beta$$
$$[h_\beta, x_\alpha] = -2x_\alpha$$
$$[h_\beta, x_{\alpha+\beta}] = 0$$
$$[h_\beta, x_{2\beta+\alpha}] = 2x_{2\beta+\alpha}$$
$$[h_\alpha, x_\beta] = -x_\beta$$
$$[h_\alpha, x_\alpha] = 2x_\alpha$$
$$[h_\alpha, x_{\alpha+\beta}] = x_{\alpha+\beta}$$

$$[x_\beta, x_\alpha] = x_{\alpha+\beta}$$
$$[x_\beta, x_{\alpha+\beta}] = 2x_{2\beta+\alpha}$$
$$[x_\beta, x_{-\alpha-\beta}] = -2x_{-\alpha}$$
$$[x_\beta, x_{-2\beta-\alpha}] = -x_{-\alpha-\beta}$$
$$[x_\alpha, x_{-\alpha-\beta}] = x_{-\beta}$$
$$[x_{\alpha+\beta}, x_{-2\beta-\alpha}] = x_{-\beta}$$
$$[h_\alpha, x_{2\beta+\alpha}] = 0.$$

7. Let $\mathsf{F} = \mathbf{C}$. Fix a Chevalley basis of L, and let L' be the $\mathbf{R}$-subspace of L spanned by the elements $\sqrt{-1}\, h_i$ $(1 \le i \le \ell)$, $x_\alpha - x_{-\alpha}$, and $\sqrt{-1}$ $(x_\alpha + x_{-\alpha})$ $(\alpha \in \Phi^+)$. Prove that these elements form a basis of L over $\mathbf{C}$ (so $L \cong L' \otimes_{\mathbf{R}} \mathbf{C}$) and that L' is closed under the bracket (so L' is a Lie algebra over $\mathbf{R}$). Show that the Killing form κ' of L' is just the restriction to L' of κ, and that κ' is *negative definite*. (L' is a "compact real form" of L, associated with a compact Lie group).

8. Let $L = \mathfrak{sl}(\ell+1, \mathsf{F})$, and let K be any field of characteristic p. If $p \nmid \ell+1$, then $L(\mathsf{K})$ is simple. If $p = 2$, $\ell = 1$, then $L(\mathsf{K})$ is solvable. If $\ell > 1$, $p|\ell+1$, then Rad $L(\mathsf{K}) = Z(L(\mathsf{K}))$ consists of the scalar matrices.

9. Prove that for L of type A_ℓ, the resulting Chevalley group $G(\mathsf{K})$ of adjoint type is isomorphic to $PSL(\ell+1, \mathsf{K}) = SL(\ell+1, \mathsf{K})$ modulo scalars (the scalars being the $\ell+1^{\text{st}}$ roots of unity in K).

10. Let L be of type G_2, K a field of characteristic 3. Prove that $L(\mathsf{K})$ has a 7-dimensional ideal M (cf. the short roots). Describe the representation of $L(\mathsf{K})$ on $L(\mathsf{K})/M$.

11. The Chevalley group $G(\mathsf{K})$ acts on $L(\mathsf{K})$ as a group of Lie algebra automorphisms.

12. Is the basis of G_2 exhibited in (19.3) a Chevalley basis?

Notes

The ideas of this section all stem from Chevalley's seminal paper [3]. Our treatment follows the lecture notes of Steinberg [2], which are the best source of information about all aspects of Chevalley groups. Carter [1], Curtis [1] have good surveys of the finite groups. An algorithm for choosing signs for the Chevalley basis is described by Samelson [1], p. 54. For an approach to the existence theorem based on a detailed (but elementary) study of signs, see Tits [1].

26. Kostant's Theorem

L, H, Φ, Δ as before. Fix also a Chevalley basis $\{x_\alpha, \alpha \in \Phi; h_i, 1 \le i \le \ell\}$ of L.

In order to construct matrix groups associated with arbitrary representations ϕ of L (not just ad), we have to work inside $\mathfrak{U}(L)$. The idea is to construct a lattice, analogous to $L(\mathbf{Z})$, in an arbitrary (finite dimensional) L-module which will be invariant under all endomorphisms $\phi(x_\alpha)^m/m!$ This construction will utilize a "$\mathbf{Z}$-form" of $\mathfrak{U}(L)$, which turns out to be just the subring with 1 of $\mathfrak{U}(L)$ generated by all $x_\alpha^m/m!$ $(\alpha \in \Phi)$.

26.1. *A combinatorial lemma.*

Recall that the binomial coefficient $\binom{n}{k} = \dfrac{n(n-1)\ldots(n-k+1)}{k!}$. If n is replaced here by an element x of any commutative, associative F-algebra (with 1), the resulting expression still makes sense and can be denoted by $\binom{x}{k}$, $k \in \mathbf{Z}^+$. The familiar identity for binomial coefficients, (*) $\binom{n+1}{k} - \binom{n}{k} = \binom{n}{k-1}$, carries over as well. As usual, we interpret $\binom{n}{k}$ to be 0 whenever k is negative, while $\binom{n}{0} = 1$.

Lemma. *Let $T_1, \ldots, T_\ell$ be indeterminates, and let $f = f(T_1, \ldots, T_\ell)$ be a polynomial over F such that $f(n_1, \ldots, n_\ell) \in \mathbf{Z}$ whenever $n_1, \ldots, n_\ell \in \mathbf{Z}$. Then f is a $\mathbf{Z}$-linear combination of the polynomials $\binom{T_1}{b_1}\binom{T_2}{b_2} \ldots \binom{T_\ell}{b_\ell}$, where $b_1, \ldots, b_\ell \in \mathbf{Z}^+$ and b_i does not exceed the degree of f viewed as a polynomial in T_i.*

Proof. First notice that the conclusion is reasonable, because

$$\binom{T_i}{b_i} = \frac{T_i(T_i-1)\ldots(T_i-b_i+1)}{b_i!}$$

does take integral values when T_i is replaced by an integer. Notice too that the indicated polynomials form an F-basis for $F[T_1, \ldots, T_\ell]$.

The proof is by induction on ℓ and on the degree of f in T_ℓ. If f is a constant polynomial, it must be an integral multiple of $1 = \binom{T_\ell}{0}$, so there is nothing to prove. In general, write $f = \sum_{k=0}^{r} f_k(T_1, \ldots, T_{\ell-1})\binom{T_\ell}{k}$, where r is the degree of f in T_ℓ and $f_k(T_1, \ldots, T_{\ell-1}) \in F[T_1, \ldots, T_{\ell-1}]$. Formally substitute $T_\ell + 1$ for T_ℓ on both sides of this equation, and subtract. The identity (*) above shows that the right side equals $\sum_{k=0}^{r} f_k(T_1, \ldots, T_{\ell-1})\binom{T_\ell}{k-1}$, while the left side is a polynomial satisfying the original hypothesis on f. Repeat this process a total of r times, until all binomial coefficients on the right become 0, except $\binom{T_\ell}{r-r} = 1$, this being the coefficient of $f_r(T_1, \ldots, T_{\ell-1})$. Now f_r satisfies the hypothesis on f (it takes integral values on $\mathbf{Z}^{\ell-1}$), but it has one fewer variable than f. By induction, f_r can be expressed in the desired fashion. Moreover, $f - f_r(T_1, \ldots, T_{\ell-1})\binom{T_\ell}{r}$ satisfies the original hypothesis on f, but is of degree $<r$ in T_ℓ, whence induction again applies to complete the proof. □

26.2. *Special case:* $\mathfrak{sl}(2, F)$

In this subsection we consider the special case $L = \mathfrak{sl}(2, F)$, with standard (Chevalley) basis x, y, h. The following lemma and its corollary amount to a proof of Kostant's Theorem in this case; this will be used below to obtain the general case.

Lemma. *If* a, $c \in \mathbf{Z}^+$, *then in* $\mathfrak{U}(L)$ *we have:*

$$\frac{x^c}{c!}\frac{y^a}{a!} = \sum_{k=0}^{\min (a,c)} \frac{y^{a-k}}{(a-k)!}\binom{h-a-c+2k}{k}\frac{x^{c-k}}{(c-k)!}.$$

Proof. If $a = 0$, or $c = 0$, the right side becomes $\dfrac{x^c}{c!}$ or $\dfrac{y^a}{a!}$ (respectively). If $a = c = 1$, the equation becomes $xy = yx + h$ (which is true). In general, we proceed by induction on a and c. First let $c = 1$, and use induction on a to obtain:

$$\frac{xy^a}{a!} = \frac{xy^{a-1}}{(a-1)!}\frac{y}{a} = \frac{y^{a-1}}{(a-1)!}\frac{xy}{a} + \frac{y^{a-2}}{(a-2)!}(h-a+2)\frac{y}{a}$$

$$= \frac{y^a x}{a!} + \frac{y^{a-1}h}{a!} + \frac{y^{a-1}h}{a(a-2)!} - \frac{2y^{a-1}}{a(a-2)!} - \frac{y^{a-1}}{a(a-3)!}$$

$$= \frac{y^a x}{a!} + \frac{y^{a-1}h}{(a-1)!} - \frac{y^{a-1}}{(a-2)!} = \frac{y^a x}{a!} + \frac{y^{a-1}}{(a-1)!}(h-a+1),$$

as desired. Using this result and induction on c, the lemma quickly follows. ∎

Corollary. *For* $b \in \mathbf{Z}^+$, $\binom{h}{b}$ *is in the subring of* $\mathfrak{U}(L)$ *generated by all* $\dfrac{x^c}{c!}$ *and* $\dfrac{y^a}{a!}$ (a, $c \in \mathbf{Z}^+$).

Proof. This is clear if $b = 0$, so we may proceed by induction on b. In the lemma, choose $a = c = b$, so the right side becomes $\binom{h}{b} + \sum_{k=0}^{b-1}\frac{y^{b-k}}{(b-k)!}$ $\binom{h-2b+2k}{k}\frac{x^{b-k}}{(b-k)!}$ (and this lies in the indicated subring of $\mathfrak{U}(L)$). Let $k < b$. The polynomial $\binom{T-2b+2k}{k}$ (T an indeterminate) clearly takes integral values at integers, so Lemma 26.1 allows us to write it as **Z**-linear combination of polynomials $\binom{T}{j}$, where $j \le k$ ($< b$). In turn, induction on b shows that each $\binom{h}{j}$ is in the indicated subring of $\mathfrak{U}(L)$, so finally $\binom{h}{b}$ is. ∎

26.3. Lemmas on commutation

We return now to the general case. Since for each $\alpha \in \Phi$, the standard basis for the three dimensional simple subalgebra S_α may be taken to be part of the chosen Chevalley basis for L, the results of (26.2) can be applied freely to situations involving only $\pm \alpha$. The main problem now is to deal with pairs of linearly independent roots.

Lemma A. *Let V, W be L-modules, with respective subgroups A, B. If A, B are invariant under all endomorphisms $\dfrac{x_\alpha^t}{t!}$ $(\alpha \in \Phi,\ t \in \mathbf{Z}^+)$, then so is $A \otimes B \subset V \otimes W$.*

Proof. Recall that $x_\alpha.(v \otimes w) = x_\alpha.v \otimes w + v \otimes x_\alpha.w$. Using the binomial expansion, we see that

$$\frac{x_\alpha^t}{t!}(v \otimes w) = \sum_{k=0}^{t}\left(\frac{x_\alpha^k}{k!}\cdot v \otimes \frac{x_\alpha^{t-k}}{(t-k)!}\cdot w\right).$$

If $v \in A$, $w \in B$, each term on the right is therefore in $A \otimes B$. □

Corollary. *Let $L(\mathbf{Z})$ (as in (25.4)) be the $\mathbf{Z}$-span of the Chevalley basis chosen for L. Then for $\alpha \in \Phi,\ t \in \mathbf{Z}^+$, $\dfrac{(ad\ x_\alpha)^t}{t!}$ leaves $L(\mathbf{Z}) \otimes L(\mathbf{Z}) \otimes \ldots \otimes L(\mathbf{Z})$ invariant.*

Proof. Use Proposition 25.5 and the lemma. □

A subset Ψ of Φ is called **closed** if $\alpha, \beta \in \Psi$, $\alpha + \beta \in \Phi$ imply that $\alpha + \beta \in \Psi$. *Examples*: Φ; Φ^+; the set of all $i\alpha + j\beta \in \Phi$ ($i, j \geq 0$; α, β linearly independent roots).

Lemma B. *Let Ψ be a closed set of roots, with $\Psi \cap -\Psi = \varnothing$. Let $\mathfrak{X}$ be the subring of $\mathfrak{U}(L)$ (with 1) generated by all $x_\alpha^t/t!$ ($\alpha \in \Psi$, $t \in \mathbf{Z}^+$). Relative to any fixed ordering of Ψ, the set of products $\prod\limits_{\alpha \in \Psi} \dfrac{x_\alpha^{t_\alpha}}{t_\alpha!}$ (written in the given order) forms a basis for the $\mathbf{Z}$-module $\mathfrak{X}$.*

Proof. It is clear that the F-span of the L_α ($\alpha \in \Psi$) is a subalgebra X of L; the PBW theorem, applied to $\mathfrak{U}(X)$, shows that the indicated products do form a basis for $\mathfrak{X}$ over F. So it will suffice to show that coefficients all lie in $\mathbf{Z}$. Let us call $\sum\limits_{\alpha \in \Psi} t_\alpha$ the *degree* of $\prod\limits_{\alpha \in \Psi} \dfrac{x_\alpha^{t_\alpha}}{t_\alpha!}$. If $x \in \mathfrak{X}$ is nonscalar, then $x = c \prod \dfrac{x_\alpha^{t_\alpha}}{t_\alpha!} + $ (terms of degree $\leq \sum t_\alpha$), where $0 \neq c \in$ F and the remaining terms of degree $t = \sum t_\alpha$ involve sequences distinct from $(\ldots t_\alpha \ldots)$. Now $\mathfrak{X}$ acts (via the adjoint representation) on $L \otimes \ldots \otimes L$ (t copies). In particular, look at $x \cdot (\underbrace{(x_{-\alpha} \otimes \ldots \otimes x_{-\alpha})}_{t_\alpha} \otimes \underbrace{(x_{-\beta} \otimes \ldots \otimes x_{-\beta})}_{t_\beta} \otimes \ldots)$, where $\Psi =$

$(\alpha, \beta, \ldots)$. What is the component of this element in $H \otimes \ldots \otimes H$ (relative to the standard PBW basis)? The first term in x, $\prod_{\alpha} \dfrac{x_{\alpha}^{t_{\alpha}}}{t_{\alpha}!}$, yields $\underbrace{c((h_{\alpha} \otimes \ldots \otimes h_{\alpha})}_{t_{\alpha}}$

$\otimes \underbrace{(h_{\beta} \otimes \ldots \otimes h_{\beta})}_{t_{\beta}} \otimes \ldots)$, by inspection. However, all other terms $\prod \dfrac{x_{\alpha}^{u_{\alpha}}}{u_{\alpha}!}$ in x, applied to the above element, yield no nonzero component in $H \otimes \ldots \otimes H$: either there are too few factors ($\sum u_{\alpha} < \sum t_{\alpha}$), or else $\sum u_{\alpha} = \sum t_{\alpha}$, but the factors are "distributed" wrong.

Now the corollary of Lemma A says that x preserves $L(\mathbf{Z}) \otimes \ldots \otimes L(\mathbf{Z})$ (t copies). Moreover, $L(\mathbf{Z})$ is independent of the choice of Δ, so we may assume that α (the first element in Ψ) is *simple*. The h_{β} (β simple) form a basis for the free $\mathbf{Z}$-module $H(\mathbf{Z}) = L(\mathbf{Z}) \cap H$, so their various tensor products form a basis for the free $\mathbf{Z}$-module $H(\mathbf{Z}) \otimes \ldots \otimes H(\mathbf{Z})$. On the other hand, we have just shown that $c((h_{\alpha} \otimes \ldots \otimes h_{\alpha}) \otimes (h_{\beta} \otimes \ldots \otimes h_{\beta}) \otimes \ldots) \in H(\mathbf{Z})$ $\otimes \ldots \otimes H(\mathbf{Z})$. This makes it clear that $c \in \mathbf{Z}$. Now the term $c \prod \dfrac{x_{\alpha}^{t_{\alpha}}}{t_{\alpha}!}$ may be subtracted from x and the argument repeated for the element $x - c \prod \dfrac{x_{\alpha}^{t_{\alpha}}}{t_{\alpha}!} \in \mathfrak{X}$. Induction on the number of terms completes the proof of the lemma. $\square$

Let us, for convenience, call any product of elements $\dfrac{x_{\alpha}^{t}}{t!}$, $\dbinom{h_i - j}{k}$ ($j, k \in \mathbf{Z}$, $t \in \mathbf{Z}^+$), in $\mathfrak{U}(L)$ a *monomial*, of *degree* equal to the sum of the various t's occurring.

Lemma C. *Let* $\alpha, \beta \in \Phi$, $k, m \in \mathbf{Z}^+$. *Then* $\dfrac{x_{\beta}^{m} x_{\alpha}^{k}}{m! \, k!}$ *is a* $\mathbf{Z}$-*linear combination of* $\dfrac{x_{\alpha}^{k} x_{\beta}^{m}}{k! \, m!}$ *along with other monomials of degree* $< k + m$.

Proof. If $\alpha = \beta$, there is nothing to prove. If $\alpha = -\beta$, this follows from Lemma 26.2. Otherwise, α, β are independent and we can apply Lemma B above to the set of roots of the form $i\alpha + j\beta$ ($i, j \geq 0$), to write the given monomial as $\mathbf{Z}$-linear combination of $\dfrac{x_{\alpha}^{k} x_{\beta}^{m}}{k! \, m!}$ and other monomials. It remains to be seen that these other monomials have degree $< k + m$. But the PBW Theorem (17.3) already assures us that $\dfrac{x_{\beta}^{m} x_{\alpha}^{k}}{m! \, k!} = \dfrac{x_{\alpha}^{k} x_{\beta}^{m}}{k! \, m!} + $ (F-linear combination of PBW basis elements of degree $< k + m$). Since we are using a PBW basis (or scalar multiples thereof), the proof is complete. $\square$

Lemma D. *Let* $\alpha, \beta \in \Phi$, $f(T) \in \mathsf{F}[T]$ (T *an indeterminate*). *Then for all* $k \in \mathbf{Z}^+$, $x_{\alpha}^{k} f(h_{\beta}) = f(h_{\beta} - k\alpha(h_{\beta})) x_{\alpha}^{k}$.

Proof. It suffices (by linearity) to prove this when f is of the form T^m; then the assertion becomes: $x_{\alpha}^{k} h_{\beta}^{m} = (h_{\beta} - k\alpha(h_{\beta}))^{m} x_{\alpha}^{k}$. If $k = 0$ or $m = 0$, this is clear. Proceed by induction on k and on m. If $k = m = 1$, we have $x_\alpha h_\beta = h_\beta x_\alpha - \alpha(h_\beta) x_\alpha = (h_\beta - \alpha(h_\beta)) x_\alpha$, as required. In general, $x_\alpha^k h_\beta^m = (x_\alpha^k h_\beta^{m-1}) h_\beta = (h_\beta - k\alpha(h_\beta))^{m-1} (x_\alpha^k h_\beta) = (h_\beta - k\alpha(h_\beta))^{m-1} (h_\beta - k\alpha(h_\beta)) x_\alpha^k$, by induction first on m and then on $k \geq 1$. $\square$

26.4. Proof of Kostant's Theorem

Fix some ordering $(\alpha_1, \ldots, \alpha_m)$ of Φ^+. Denote m-tuples or ℓ-tuples of nonnegative integers by $A = (a_1, \ldots, a_m)$, $B = (b_1, \ldots, b_\ell)$, $C = (c_1, \ldots, c_m)$. Then define elements of $\mathfrak{U}(L)$ as follows:

$$f_A = \frac{x_{-\alpha_1}^{a_1}}{a_1!} \cdots \frac{x_{-\alpha_m}^{a_m}}{a_m!}$$

$$h_B = \binom{h_1}{b_1} \cdots \binom{h_\ell}{b_\ell}$$

$$e_C = \frac{x_{\alpha_1}^{c_1}}{c_1!} \cdots \frac{x_{\alpha_m}^{c_m}}{c_m!}.$$

Notice that the various h_B form a basis over F for $\mathfrak{U}(H)$: this was essentially remarked in (26.1), in connection with polynomials. Combined with the PBW Theorem, this shows that the various elements $f_A h_B e_C$ form an F-basis of $\mathfrak{U}(L)$.

Theorem (Kostant). Let $\mathfrak{U}(L)_\mathbf{Z}$ be the subring of $\mathfrak{U}(L)$ (with 1) generated by all $x_\alpha^t/t!$ ($\alpha \in \Phi$, $t \in \mathbf{Z}^+$). Let $\mathfrak{B}$ be the lattice in $\mathfrak{U}(L)$ with $\mathbf{Z}$-basis consisting of all $f_A h_B e_C$. Then $\mathfrak{B} = \mathfrak{U}(L)_\mathbf{Z}$.

Proof. This is just a matter of fitting the pieces together. First of all, each $\binom{h_i}{b_i} \in \mathfrak{U}(L)_\mathbf{Z}$, thanks to the Corollary of Lemma 26.2. Therefore $\mathfrak{B} \subset \mathfrak{U}(L)_\mathbf{Z}$.

The reverse inclusion is harder to prove. It will suffice to show that each "monomial" (as defined in (26.3)) lies in $\mathfrak{B}$, since these span the $\mathbf{Z}$-module $\mathfrak{U}(L)_\mathbf{Z}$. For this we use induction on "degree". Monomials of degree 0 involve only factors of the form $\binom{h_i - j}{k}$, and lie in $\mathfrak{B}$ thanks to Lemma 26.1. In general, Lemmas C and D of (26.3) (along with the induction hypothesis) permit us to write a monomial as $\mathbf{Z}$-linear combination of other monomials for which factors involving $x_{-\alpha}$'s, h's, and x_α's come in the prescribed order. The identity $\frac{T^k}{k!} \frac{T^m}{m!} = \binom{m+k}{m} \frac{T^{k+m}}{(k+m)!}$ further insures that each $x_{\pm\alpha}$ will appear in at most one term of each resulting monomial. Now Lemma 26.1 and Lemma D of (26.3) allow us to complete the proof. $\square$

Exercises

1. Let $L = \mathfrak{sl}(2, F)$. Let $(v_0, v_1, \ldots, v_m)$ be the basis constructed in (7.2) for the irreducible L-module $V(m)$ of highest weight m. Prove that the Z-span of this basis is invariant under $\mathfrak{U}(L)_Z$. Let $(w_0, w_1, \ldots, w_m)$ be the basis of $V(m)$ used in (22.2). Show that the Z-span of the w_i is *not* invariant under $\mathfrak{U}(L)_Z$.

2. Let $\lambda \in \Lambda^+ \subset H^*$ be a dominant integral linear function, and recall the module $Z(\lambda)$ of (20.3), with irreducible quotient $V(\lambda) = Z(\lambda)/Y(\lambda)$. Show that the multiplicity of a weight μ of $V(\lambda)$ can be effectively computed as follows, thanks to Kostant's Theorem: If v^+ is a maximal vector of $Z(\lambda)$, then the various $f_A.v^+$ for which $\sum a_i \alpha_i = \lambda - \mu$ form an F-basis of the weight space for μ in $Z(\lambda)$. (Cf. Lemma D of (24.1).) In turn if $\sum a_i \alpha_i = \sum c_i \alpha_i$, then $e_C f_A.v^+$ is an integral multiple $n_{CA} v^+$. This yields a $d \times d$ integral matrix (n_{CA}) (d = multiplicity of μ in $Z(\lambda)$), whose *rank* = $m_\lambda(\mu)$. (Cf. Exercise 20.9). Moreover, this integral matrix is computable once the Chevalley basis structure constants are known. Carry out a calculation of this kind for type A_2, taking $\lambda - \mu$ small.

Notes

Theorem 26.4 appears in Kostant [2]; here we reproduce the proof in Steinberg [2]. For related material, from a "schematic" point of view, cf. Chevalley [4], Borel [2]. Exercise 2 is based on Burgoyne [1].

27. Admissible lattices

The notation is that of §26. Using Kostant's Theorem, we shall construct an "admissible" lattice in an arbitrary finite dimensional L-module and describe its stabilizer in L. Reduction modulo a prime then yields linear groups and linear Lie algebras over an arbitrary field of prime characteristic, generalizing the construction of Chevalley groups and Chevalley algebras given in §25.

27.1. Existence of admissible lattices

It follows from Kostant's Theorem (or the lemmas preceding it) that if $N^+ = \coprod_{\alpha > 0} L_\alpha$, $N^- = \coprod_{\alpha < 0} L_\alpha$, then each of $\mathfrak{U}(N^-)$, $\mathfrak{U}(H)$, $\mathfrak{U}(N^+)$ has a "Z-form" with Z-basis consisting of all f_A, h_B, e_C (respectively). We call these subrings $\mathfrak{U}_Z^-$, $\mathfrak{U}_Z^0$, $\mathfrak{U}_Z^+$, so $\mathfrak{U}_Z$ $(= \mathfrak{U}(L)_Z)$ equals $\mathfrak{U}_Z^- \mathfrak{U}_Z^0 \mathfrak{U}_Z^+$.

One further preliminary: We have defined a *lattice* M in a finite dimensional vector space V over F to be the Z-span of a basis of V (over F). Since char $F = 0$, a finitely generated $\mathbf{Z}$-submodule of V is automatically a free $\mathbf{Z}$-module of finite rank. Therefore, a lattice in V may be characterized as a finitely generated subgroup of V which spans V over F and has $\mathbf{Z}$-rank $\leq \dim_F V$.

Lemma. *Let* $d \in \mathbf{Z}^\ell$, $S \subset \mathbf{Z}^\ell$ *a finite set not containing* d. *Then there exists a polynomial* $f(T_1, \ldots, T_\ell)$ *over* F *such that* $f(\mathbf{Z}^\ell) \subset \mathbf{Z}, f(d) = 1$, *and* $f(S) = 0$.

Proof. Say $d = (d_1, \ldots, d_\ell)$. If $k \in \mathbf{Z}^+$, set $f_k(T_1, \ldots, T_\ell) = \prod_{i=1}^\ell$ $\binom{T_i - d_i + k}{k} \binom{-T_i + d_i + k}{k}$, so $f_k(\mathbf{Z}^\ell) \subset \mathbf{Z}$ (cf. Lemma 26.1), $f_k(d) = 1$. In the "box" in $\mathbf{Z}^\ell$ centered at d, with edge $2k$, f_k evidently takes the value 0 except at d. So it suffices to choose k large enough for this "box" to capture the finite set S, and take $f = f_k$. $\square$

Theorem. *Let* V *be a finite dimensional* L-*module. Then*: (a) *Any subgroup of* V *invariant under* $\mathfrak{U}_\mathbf{Z}$ *is the direct sum of its intersections with the weight spaces of* V.
(b) V *contains a lattice which is invariant under* $\mathfrak{U}_\mathbf{Z}$.

Proof. (a) Let M be a subgroup of V stable under $\mathfrak{U}_\mathbf{Z}$. For each weight μ of V, set $d(\mu) = (\mu(h_1), \ldots, \mu(h_\ell)) \in \mathbf{Z}^\ell$. Fix an arbitrary weight λ of V. The preceding lemma then yields a polynomial f over F in ℓ variables, such that $f(\mathbf{Z}^\ell) \subset \mathbf{Z}, f(d(\lambda)) = 1, f(d(\mu)) = 0$ for $\mu \neq \lambda$ in $\Pi(V)$. Set $u = f(h_1, \ldots, h_\ell)$. Thanks to Lemma 26.1, $u \in \mathfrak{U}_\mathbf{Z}^0$. Evidently u acts on V as projection onto V_λ. In particular, if $v \in M$, its V_λ-component $u.v$ also lies in M.

(b) In view of Weyl's Theorem on complete reducibility, we may assume that $V = V(\lambda)$ $(\lambda \in \Lambda^+)$, i.e., that V is *irreducible*. Let $v^+ \in V$ be a maximal vector (of weight λ), and set $M = \mathfrak{U}_\mathbf{Z}^-.v^+$. Since all elements except 1 in the $\mathbf{Z}$-basis $\{e_C\}$ of $\mathfrak{U}_\mathbf{Z}^+$ kill v^+, we have $\mathfrak{U}_\mathbf{Z}^+.v^+ = \mathbf{Z}v^+$. Also, $\mathfrak{U}_\mathbf{Z}^0.v^+ = \mathbf{Z}v^+$, because $\binom{h_i}{b_i}$ acts on v^+ as scalar multiplication by the integer

$$\frac{\lambda(h_i)(\lambda(h_i) - 1) \ldots (\lambda(h_i) - b_i + 1)}{b_i!}.$$

In other words, $\mathfrak{U}_\mathbf{Z}.v^+ = \mathfrak{U}_\mathbf{Z}^- \mathfrak{U}_\mathbf{Z}^0 \mathfrak{U}_\mathbf{Z}^+.v^+ = \mathfrak{U}_\mathbf{Z}^-.(\mathbf{Z}v^+) = M$, so M is *invariant under* $\mathfrak{U}_\mathbf{Z}$. The argument also shows that $M \cap V_\lambda = \mathbf{Z}v^+$. We know that all but finitely many of the f_A kill v^+, so M is *finitely generated*. Moreover, since $\mathfrak{U}_\mathbf{Z}^-$ contains an F-basis of $\mathfrak{U}(N^-)$, while $\mathfrak{U}(N^-).v^+ = V$, M spans V over F.

It remains to be seen that the $\mathbf{Z}$-rank of M does not exceed $\dim_F V$. Suppose the contrary, and let r be the smallest number of vectors in M which are free over $\mathbf{Z}$ but linearly dependent over F. Say $\sum_{i=1}^r a_i v_i = 0$ ($a_i \in F, 0 \neq v_i \in M$). For some $u \in \mathfrak{U}_\mathbf{Z}$, $u.v_1$ must have nonzero V_λ-component: otherwise v_1 would generate a nonzero proper $\mathfrak{U}(L)$-submodule of the irreducible module V. On the other hand, the V_λ-component of each $u.v_i$ ($1 \leq i \leq r$) lies in M

(by part (a)), hence is an integral multiple of v^+, say $m_i v^+$ (because $M \cap V_\lambda = \mathbf{Z} v^+$). So $\sum a_i v_i = 0$ implies $\sum a_i (u.v_i) = 0$, and in turn $\sum a_i m_i = 0$ (but $m_1 \neq 0$). Therefore, $0 = m_1 \left(\sum_{i=1}^{r} a_i v_i \right) - \left(\sum_{i=1}^{r} a_i m_i \right) v_1 = \sum_{i=2}^{r} a_i (m_1 v_i - m_i v_1)$. The vectors $m_1 v_i - m_i v_1$ $(2 \leq i \leq r)$ lie in M and are evidently free over $\mathbf{Z}$ but linearly dependent over F. This contradicts the minimality of r, and proves that M is a lattice in V, stable under $\mathfrak{U}_\mathbf{Z}$. $\square$

A lattice M in a finite dimensional L-module V which is invariant under $\mathfrak{U}_\mathbf{Z}$ is called **admissible**. Part (b) of the theorem asserts the existence of such a lattice; actually, the proof shows how to construct one (the smallest possible one containing a given maximal vector, if V is irreducible). Part (a) implies that $M = \coprod_{\mu \in \Pi(V)} (M \cap V_\mu)$. Of course, when V is L itself (for the representation ad), the $\mathbf{Z}$-span of a Chevalley basis has already been seen to be an admissible lattice (25.5).

27.2. Stabilizer of an admissible lattice

Let V be a finite dimensional L-module. To avoid trivialities we assume that V is *faithful* (in other words, we discard those simple ideals of L which act trivially on V). Then it is easy to see that the $\mathbf{Z}$-span of $\Pi(V)$, call it $\Lambda(V)$, lies between Λ and the root lattice Λ_r (Exercise 21.5).

Using Theorem 27.1, choose an admissible lattice M in V, and let L_V be its stabilizer in L, $H_V = H \cap L_V$. (It will be shown below that L_V depends only on V, not on the choice of M, so the notation is unambiguous). Obviously $L(\mathbf{Z}) \subset L_V$, and L_V is closed under the bracket. To say that $h \in H$ leaves M invariant is just to say that $\lambda(h) \in \mathbf{Z}$ for all $\lambda \in \Pi(V)$ (or $\Lambda(V)$), in view of part (a) of Theorem 27.1. This shows that the lattice inclusions $\Lambda \supset \Lambda(V) \supset \Lambda_r$ induce reverse inclusions $H(\mathbf{Z}) \subset H_V \subset H_0$, where $H_0 = \{h \in H | \lambda(h) \in \mathbf{Z}$ for all $\lambda \in \Lambda_r\}$ and $H(\mathbf{Z}) = H \cap L(\mathbf{Z})$ $(=\mathbf{Z}$-span of all h_α, $\alpha \in \Phi)$. In particular, H_V is a lattice in H. Our aim is to show that L_V is an admissible lattice in L. The following general lemma is a first step. (It could be formulated as a fact about associative algebras, but we need only a special case of it.)

Lemma. *If $u \in \mathfrak{U}(L)$, $x \in L$, then*

$$\frac{(ad\ x)^n}{n!}(u) - \sum_{i=0}^{n}(-1)^i \frac{x^{n-i}}{(n-i)!} u \frac{x^i}{i!}$$

in $\mathfrak{U}(L)$.

Proof. For $n = 1$, this reads ad $x\,(u) = xu - ux$, which is true by definition. Use induction on n:

$$\frac{(ad\ x)^n}{n!}(u) = \frac{ad\ x}{n} \left(\sum_{i=0}^{n-1}(-1)^i \frac{x^{n-1-i}}{(n-1-i)!} u \frac{x^i}{i!} \right)$$

$$= \left(\sum_{i=0}^{n-1}(-1)^i \frac{x^{n-i}}{n(n-1-i)!} u \frac{x^i}{i!} \right) - \left(\sum_{i=0}^{n-1}(-1)^i \frac{x^{n-1-i}}{(n-1-i)!} u \frac{x^{i+1}}{i!n} \right).$$

After the change of index $i \mapsto i-1$, the second sum reads:

$$- \sum_{i=1}^{n} (-1)^i \frac{x^{n-i}}{(n-i)!} u \frac{x^i}{(i-1)!n} .$$

For $0 < i < n$, the ith term combines with the ith term of the first sum to yield

$$(-1)^i x^{n-i} u x^i \left(\frac{1}{n(n-1-i)!i!} + \frac{1}{n(n-i)!(i-1)!} \right) .$$

But the quantity in parentheses is just $\frac{1}{(n-i)!i!}$. The 0th and nth terms are

$\frac{x^n}{n!} u$ and $(-1)^n u \frac{x^n}{n!}$, as required. $\square$

Proposition. L_V *is an admissible lattice in the L-module L. Moreover,* $L_V = H_V + \coprod_{\alpha \in \Phi} \mathbf{Z} x_\alpha$; *therefore,* L_V *depends only on V (or just $\Lambda(V)$), not on the choice of M.*

Proof. We know that $L(\mathbf{Z}) = H(\mathbf{Z}) + \coprod \mathbf{Z} x_\alpha \subset L_V$, and clearly $H_V \subset L_V$. On the other hand, the preceding lemma guarantees that L_V is invariant under all $(\text{ad } x_\alpha)^m/m!$ (hence under $\mathfrak{U}_\mathbf{Z}$). This allows us to write L_V as the sum of its intersections with H and the L_α (part (a) of Theorem 27.1), so $L_V = H_V + \coprod (L_V \cap L_\alpha)$, with $\mathbf{Z} x_\alpha \subset L_V \cap L_\alpha$. The proposition will follow at once if we prove that this last inclusion is an equality for each $\alpha \in \Phi$.

Consider the linear map $\phi: L_\alpha \to H$ defined by the rule $x \mapsto [x_{-\alpha} x]$ ($=$ multiple of $h_{-\alpha}$). This is injective, since $\dim L_\alpha = 1$ and $[x_{-\alpha} L_\alpha] \neq 0$. The restriction of ϕ to $L_V \cap L_\alpha$ has image in H_V (since L_V is closed under the bracket, and $H_V = L_V \cap H$). So this image is *infinite cyclic*, H_V being a lattice in H (free abelian of rank ℓ). Say $\frac{1}{n} x_\alpha$ ($n \in \mathbf{Z}^+$) is a generator of the infinite cyclic group $L_V \cap L_\alpha$: $x_\alpha \subset L_V \cap L_\alpha$ must of course be an *integral* multiple of the generator. Then $\frac{(\text{ad } x_{-\alpha})^2}{2!} \left(\frac{x_\alpha}{n} \right) = \frac{x_{-\alpha}}{n} \in L_V$ (because L_V is stable under $\mathfrak{U}_\mathbf{Z}$), and in turn $-\left(\text{ad } \frac{x_\alpha}{n} \right)^2 \left(\frac{x_{-\alpha}}{n} \right) = \frac{2 x_\alpha}{n^3} \in L_V$ (because L_V is closed under the bracket). But then $\frac{2}{n^3} \in \frac{1}{n} \mathbf{Z}$, forcing $n = 1$. This shows that $L_V \cap L_\alpha = \mathbf{Z} x_\alpha$, as required. $\square$

As an example, consider $L = \mathfrak{sl}(2, \mathsf{F})$, with standard basis (x, y, h). For the usual two dimensional representation of L, $V = \mathsf{F}^2$, the basis $\{(1, 0), (0, 1)\}$ obviously spans an admissible lattice, with $L_V = \mathbf{Z} h + \mathbf{Z} x + \mathbf{Z} y$ ($= L(\mathbf{Z})$). On the other hand, taking $L(\mathbf{Z})$ as admissible lattice in L (for the 3-dimensional representation ad), we find $L_V = \mathbf{Z} \left(\frac{h}{2} \right) + \mathbf{Z} x + \mathbf{Z} y$. These extreme cases correspond to the two possible weight lattices Λ, Λ_r for the root system A_1 (Exercise 4).

27.3. *Variation of admissible lattice*

Let $V = V(\lambda)$ ($\lambda \in \Lambda^+$) be an irreducible L-module. What are the possible choices of admissible lattice in V? In view of Theorem 27.1(a), such a lattice M must contain a maximal vector v^+, hence must include the admissible lattice $\mathfrak{U}_{\mathbf{Z}}^- . v^+ = \mathfrak{U}_{\mathbf{Z}} . v^+$ utilized in the proof of Theorem 27.1(b). Since v^+ is uniquely determined by V, up to a scalar multiple, we can keep v^+ fixed throughout the discussion and denote this minimal admissible lattice $\mathfrak{U}_{\mathbf{Z}} . v^+$ by $M_{\min}$. We would be satisfied now to know *which other admissible lattices intersect V in $\mathbf{Z}v^+$.*

Recall the notion of dual (or contragredient) module: V^* is the vector space dual of V, with L acting by the rule $(x.f)(w) = -f(x.w)$ ($x \in L$, $w \in V$, $f \in V^*$). If X is a subspace of V^* invariant under L, then the corresponding subspace $X^\perp$ of V is easily seen to be invariant under L, so in particular V^* is again irreducible. In fact, we can specify its highest weight (Exercise 21.6): Let $\sigma \in \mathscr{W}$ send Δ to $-\Delta$ (hence Φ^+ to Φ^-), and let $w \in V$ be a nonzero vector of weight $\sigma\lambda$. So w is a "minimal vector", killed by all $x_{-\alpha}$. Of course, dim $V_{\sigma\lambda} = $ dim $V_\lambda = 1$, so w is essentially unique. Relative to a basis of V consisting of w along with other weight vectors, take f^+ to be the linear function dual to w. Then f^+ is a maximal vector of V^*, of weight $-\sigma\lambda$, and $V^* \cong V(-\sigma\lambda)$.

Now let M be an admissible lattice in V. Define $M^* = \{f \in V^* | f(M) \subset \mathbf{Z}\}$. If M is the $\mathbf{Z}$-span of a certain basis of V, then M^* is evidently the $\mathbf{Z}$-span of the dual basis. In particular, M^* is a lattice. It is even *admissible*: $v \in M$, $f \in M^*$ implies that $((x_\alpha^m/m!).f)(v) = \pm f((x_\alpha^m/m!).v) \in \mathbf{Z}$. It is also clear that an inclusion $M_1 \subset M_2$ of admissible lattices in V induces a reverse inclusion $M_1^* \supset M_2^*$.

Assume now that v^+ (hence $M_{\min}$) has been fixed, as above. Then there is a canonical way to choose a minimal vector in $V_{\sigma\lambda} \cap M_{\min}$. Just notice that the Weyl reflections constructed in step (5) of the proof of Theorem 21.2 are transformations of V representing elements of $\mathfrak{U}_{\mathbf{Z}}$; in particular, σ maps $\mathbf{Z}v^+$ onto $M_{\min} \cap V_{\sigma\lambda}$. We may define w to be the image of v^+, and then take $f^+ \in V_{-\sigma\lambda}^*$ as above (part of a dual basis relative to a basis of $M_{\min}$). It follows immediately that $M_{\min}^* \cap V_{-\sigma\lambda}^* = \mathbf{Z}f^+$.

Now let M be any admissible lattice in V which intersects V_λ precisely in $\mathbf{Z}v^+$. The preceding argument shows that M intersects $V_{\sigma\lambda}$ precisely in $\mathbf{Z}w$, hence also that M^* intersects $V_{-\sigma\lambda}^*$ precisely in $\mathbf{Z}f^+$. Therefore, as M varies over the collection of all admissible lattices in V of the type indicated, M^* ranges over the analogous collection of admissible lattices in V^*, but with inclusions reversed. This shows that M^* has an upper as well as lower "bound", so the same is true for M. (For ad, we already proved this in another way, by considering dual lattices for H alone.)

Proposition. *Let $V = V(\lambda)$, $\lambda \in \Lambda^+$, with maximal vector v^+.*

(a) Each admissible lattice intersecting V_λ in $\mathbf{Z}v^+$ includes $M_{\min} = \mathfrak{U}_{\mathbf{Z}}.v^+$.

(b) Each admissible lattice intersecting V_λ in $\mathbf{Z}v^+$ is included in $M_{\max}$, the lattice dual to a suitable $(M^*)_{\min}$ in V^*. $\square$

27.4. Passage to an arbitrary field

Let $\mathbf{F}_p$ be the prime field of characteristic p, K an extension field of $\mathbf{F}_p$. Let V be a faithful L-module; then the weights of V span a lattice between Λ_r and Λ, which we have denoted $\Lambda(V)$. Choose an admissible lattice M in V, with stabilizer $L_V = H_V + \coprod_{\alpha \in \Phi} \mathbf{Z}x_\alpha$ in L, $H_V = \{h \in H | \lambda(h) \in \mathbf{Z}$ for all $\lambda \in \Lambda(V)\}$ (27.2).

Let $V(\mathrm{K}) = M \otimes_{\mathbf{Z}} \mathrm{K}$, $L_V(\mathrm{K}) = L_V \otimes_{\mathbf{Z}} \mathrm{K}$. Since L_V is isomorphic to a subgroup of End M (and is closed under the bracket), $L_V(\mathrm{K})$ may be identified with a Lie subalgebra of $\mathfrak{gl}(V(\mathrm{K}))$ ($=$ End $V(\mathrm{K})$). Moreover, the inclusion $L(\mathbf{Z}) \to L_V$ induces a Lie algebra homomorphism $L(\mathrm{K}) \to L_V(\mathrm{K})$, which is injective on $\coprod \mathrm{K}x_\alpha$ but may have a nonzero kernel in $H(\mathbf{Z}) \otimes_{\mathbf{Z}} \mathrm{K} = H(\mathrm{K})$. To see how this works, recall the discussion of $\mathfrak{sl}(2, \mathrm{F})$ in (27.2). If $p = 2$, then $h \otimes 1$ in $L(\mathrm{K})$ is sent to $2\left(\dfrac{h}{2} \otimes 1\right) = 0$ in $L_V(\mathrm{K})$ when $V = L$ (adjoint representation). Moreover, the multiplication in $L(\mathrm{K})$ differs from that in $L_V(\mathrm{K})$: e.g., in the former $[hx] = 2x = 0$, while in the latter $\left[\dfrac{h}{2}x\right] = x \neq 0$. On the other hand, when $p > 2$, $L(\mathrm{K}) \to L_V(\mathrm{K})$ is an isomorphism (for either choice of weight lattice $\Lambda(V) = \Lambda$ or Λ_r): Exercise 5.

The preceding discussion shows that each faithful L-module V gives rise, via the homomorphism $L(\mathrm{K}) \to L_V(\mathrm{K})$, to a module $V(\mathrm{K})$ for $L(\mathrm{K})$, which is occasionally not faithful (when $L(\mathrm{K})$ has an ideal, necessarily central, included in $H(\mathrm{K})$). It must be emphasized that (in spite of the notation) *all of this depends on the choice of the admissible lattice M in V.*

What is the analogue here of the Chevalley group $G(\mathrm{K})$ constructed in (25.5)? Since M is stable under the endomorphism $\dfrac{x_\alpha^t}{t!}$ (i.e., under $\dfrac{\phi(x_\alpha)^t}{t!}$, if ϕ is the representation of L in question), $V(\mathrm{K})$ is stable under the corresponding endomorphism, which we call $x_{\alpha, t}$ (where $x_{\alpha, 0} = 1$). Notice that for $t < p$, $x_{\alpha, t}$ acts just like $\dfrac{(x_\alpha \otimes 1)^t}{t!}$, whereas for $t \geq p$ this notation no longer has a meaning. At any rate, $x_{\alpha, t} = 0$ for large enough t, so we may write $\theta_\alpha(1) = \sum_{t=0}^{\infty} x_{\alpha, t} \in$ End $(V(\mathrm{K}))$. Actually, it is clear that $\theta_\alpha(1)$ has determinant 1, so it belongs to $SL(V(\mathrm{K}))$. More generally, we can define automorphisms $\theta_\alpha(c)$ of $V(\mathrm{K})$ by forming $\sum_{t=0}^{\infty} \dfrac{(Tx_\alpha)^t}{t!}$ and then specializing the indeterminate T to $c \in \mathrm{K}$. The group $G_V(\mathrm{K})$ generated by all $\theta_\alpha(c)$ ($\alpha \in \Phi$,

$c \in K$) is called a **Chevalley group of type $\Lambda(V)$, adjoint** if $\Lambda(V) = \Lambda_r$, **universal** if $\Lambda(V) = \Lambda$. As before, $G_V(K)$ actually depends on the choice of M.

27.5. Survey of related results

The constructions just described raise many questions, not all of which are settled. To give the reader some idea of what is known, we list now a number of results (without proof):

(1) Up to isomorphism, $G_V(K)$ and $L_V(K)$ depend on the weight lattice $\Lambda(V)$ but not on V itself or on the choice of M. (M does, however, affect the *action* of $G_V(K)$, $L_V(K)$ on $V(K)$.) If $\Lambda(V) \supset \Lambda(W)$, there exist canonical homomorphisms $G_V(K) \rightarrow G_W(K)$, $L_W(K) \rightarrow L_V(K)$. In particular, the Chevalley groups of universal type ($\Lambda(V) = \Lambda$) "cover" all others, while those of adjoint type are "covered" by all others.

(2) Let $V = V(\lambda)$, $\lambda \in \Lambda^+$, $M = M_{\min}$. Then $V(K)$ is a cyclic module for $G_V(K)$, generated by a vector $v \otimes 1$, $v \in M \cap V_\lambda$. As a result, $V(K)$ has a unique maximal $G_V(K)$-submodule, hence a unique irreducible homomorphic image (of "highest weight" λ). On the other hand, if $M = M_{\max}$, then $V(K)$ has a unique irreducible submodule, of "highest weight" λ.

(3) When $\lambda \in \Lambda^+$ satisfies $0 \leq \lambda(h_i) < p(1 \leq i \leq \ell)$, $p = \text{char } K$, then the assertions in (2) also hold true for $L_V(K)$ (or $L(K)$) in place of $G_V(K)$; the resulting p^ℓ irreducible modules are inequivalent and exhaust the (isomorphism classes of) irreducible "restricted" $L(K)$-modules.

(4) The composition factors of $V(K)$, viewed as module for either $G_V(K)$ or $L_V(K)$, are independent of the choice of admissible lattice.

Exercises

1. If M is an admissible lattice in V, then $M \cap V_\mu$ is a lattice in V_μ for each weight μ of V.

2. Prove that each admissible lattice in L which includes $L(\mathbf{Z})$ and is closed under the bracket has the form L_V. [Imitate the proof of Proposition 27.2; cf. Exercise 21.5.]

3. If M (resp. N) is an admissible lattice in V (resp. W), then $M \otimes N$ is an admissible lattice in $V \otimes W$ (cf. Lemma 26.3A). Use this fact, and the identification (as L-modules) of $V^* \otimes V$ with End V (6.1), to prove that L_V is stable under all (ad $x_a)^m/m!$ in Proposition 27.2 (without using Lemma 27.2).

In the following exercises, $L = \mathfrak{sl}(2, F)$, and weights are identified with integers.

4. Let $V = V(\lambda)$, $\lambda \in \Lambda^+$. Prove that $L_V = L(\mathbf{Z})$ when λ is odd, while $L_V = \mathbf{Z}\left(\dfrac{h}{2}\right) + \mathbf{Z}x + \mathbf{Z}y$ when λ is even.

5. If char $K > 2$, prove that $L(K) \rightarrow L_V(K)$ is an isomorphism for any choice of V.

6. Let $V = V(\lambda)$, $\lambda \in \Lambda^+$. Prove that $G_V(\mathsf{K}) \cong SL(2, \mathsf{K})$ when $\Lambda(V) = \Lambda$, $PSL(2, \mathsf{K})$ when $\Lambda(V) = \Lambda_r$.

7. If $0 \leq \lambda < \text{char } \mathsf{K}$, $V = V(\lambda)$, prove that $V(\mathsf{K})$ is irreducible as $L(\mathsf{K})$-module.

8. Fix $\lambda \in \Lambda^+$, let $V = V(\lambda)$. Fix a **Z**-basis $(e_0, \ldots, e_\lambda)$ for an admissible lattice M in V, consisting of weight vectors e_i of weight $\lambda - 2i$ (cf. Theorem 27.1(a)). (Weight spaces being one dimensional, the e_i are unique up to scalar multiples.) Let $x.e_i = c_i e_{i-1}$ $(c_0 = 1)$, $y.e_i = d_i e_{i+1}$ $(d_\lambda = 1)$. Show that (up to signs) the two sequences of integers $(c_0, \ldots, c_\lambda)$ and $(d_\lambda, \ldots, d_0)$ are identical, and that (up to signs) the sets $\{c_1, \ldots, c_\lambda\}$ and $\{1, 2, \ldots, \lambda\}$ coincide. Finally, prove that these conditions on the c_i, d_i are enough to define an admissible lattice in V.

Notes

Much of this material is taken from Steinberg [2], cf. also Borel [2]. For some recent work, consult Burgoyne [1], Burgoyne, Williamson [1], Humphreys [1], [2], Wong [1].

References

V. K. AGRAWALA, J. G. F. BELINFANTE, [1] Weight diagrams for Lie group representations: A computer implementation of Freudenthal's algorithm in ALGOL and FORTRAN, *BIT* **9**, 301-314 (1969).

J.-P. ANTOINE, D. SPEISER, [1] Characters of irreducible representations of the simple groups, I., *J. Mathematical Physics* **5**, 1226-1234, II., Ibid., 1560-1572 (1964).

D. W. BARNES, [1] On Cartan subalgebras of Lie algebras, *Math. Z.* **101**, 350-355 (1967).

R. E. BECK, B. KOLMAN, [1] A computer implementation of Freudenthal's multiplicity formula, *Indag. Math.*, to appear

I. N. BERNSTEIN, I. M. GEL'FAND, S. I. GEL'FAND, [1] Structure of representations generated by vectors of highest weight, *Functional Anal. Appl.* **5**, 1-8 (1971).

[2] Differential operators on principal affine spaces and investigation of $\mathfrak{g}$-modules, to appear.

A. BOREL, [1] *Linear Algebraic Groups*, New York: W. A. Benjamin 1969.

[2] Properties and linear representations of Chevalley groups, Seminar on Algebraic Groups and Related Finite Groups, *Lecture Notes in Math.* **131**, Berlin-Heidelberg-New York: Springer-Verlag 1970.

N. BOURBAKI, [1] *Groupes et algèbres de Lie*, Chap. 1, Paris: Hermann 1960.

[2] *Groupes et algèbres de Lie*, Chap. 4-6, Paris: Hermann 1968.

N. BURGOYNE, [1] Modular representations of some finite groups, Representation Theory of Finite Groups and Related Topics, *Proc. Symp. Pure Math. XXI* Providence: Amer. Math. Soc. 1971.

N. BURGOYNE, C. WILLIAMSON [1] Some computations involving simple Lie algebras, *Proc. 2nd Symp. Symbolic & Alg. Manipulation*, ed. S. R. Petrick, New York: Assn. Computing Machinery 1971.

R. W. CARTER, [1] Simple groups and simple Lie algebras, *J. London Math. Soc.* **40**, 193-240 (1965).

P. CARTIER, [1] On H. Weyl's character formula, *Bull. Amer. Math. Soc.* **67**, 228-230 (1961).

C. CHEVALLEY, [1] *Théorie des Groupes de Lie, Tome II*, Paris: Hermann 1951.

[2] *Théorie des Groupes de Lie, Tome III*, Paris: Hermann 1955.

[3] Sur certains groupes simples, *Tôhoku Math. J.* (2) **7**, 14-66 (1955).

[4] Certain schémas de groupes semi-simples, *Sém. Bourbaki* 1960-61, *Exp.* 219, New York: W. A. Benjamin 1966.

C. W. CURTIS, [1] Chevalley groups and related topics, *Finite Simple Groups*, ed., M. B. Powell, G. Higman, London-New York: Academic Press 1971.

H. FREUDENTHAL, [1] Zur Berechnung der Charaktere der halbeinfachen Lieschen Gruppen, I., *Indag. Math.* **16**, 369-376 (1954); II., Ibid., 487-491; III., Ibid **18**, 511-514 (1956).

H. FREUDENTHAL, H. DE VRIES, [1] *Linear Lie Groups*, London-New York: Academic Press 1969.

HARISH-CHANDRA, [1] Some applications of the universal enveloping algebra of a semi-simple Lie algebra, *Trans. Amer. Math. Soc.* **70**, 28-99 (1951).

J. E. HUMPHREYS [1] Modular representations of classical Lie algebras and semisimple groups, *J. Algebra* **19**, 51-79 (1971).

[2] Projective modules for $SL(2, q)$, *J. Algebra*, to appear.

N. JACOBSON, [1] *Lie Algebras*, New York-London: Wiley Interscience 1962.

[2] *Exceptional Lie Algebras*, New York: Marcel Dekker 1971.

I. KAPLANSKY, [1] *Lie Algebras and Locally Compact Groups*, Chicago-London: U. Chicago Press 1971.

A. U. KLIMYK, [1] Decomposition of a tensor product of irreducible representations of a semisimple Lie algebra into a direct sum of irreducible representations, *Amer. Math. Soc. Translations*, Series 2, vol. 76, Providence: Amer. Math. Soc. 1968.

B. KOLMAN, J. G. F. BELINFANTE, [1] *A Survey of Lie Groups and Lie Algebras with Computational Methods and Applications*, Philadelphia: SIAM 1972.

B. KOSTANT, [1] A formula for the multiplicity of a weight, *Trans. Amer. Math. Soc.* **93**, 53-73 (1959).

[2] Groups over Z, Algebraic Groups and Discontinuous Subgroups, *Proc. Symp. Pure Math.* IX, Providence: Amer. Math. Soc. 1966.

M. I. KRUSEMEYER, [1] Determining multiplicities of dominant weights in irreducible Lie algebra representations using a computer, *BIT* **11**, 310-316 (1971).

F. W. LEMIRE, [1] Existence of weight space decompositions for irreducible representations of simple Lie algebras, *Canad. Math. Bull.* **14**, 113-115 (1971).

R. D. POLLACK, [1] Introduction to Lie Algebras, notes by G. Edwards, Queen's Papers in Pure and Applied Math., No. 23, Kingston, Ont.: Queen's University 1969.

H. SAMELSON, [1] *Notes on Lie Algebras*, Van Nostrand Reinhold Mathematical Studies No. 23, New York: Van Nostrand Reinhold 1969.

R. D. SCHAFER, [1] *An Introduction to Nonassociative Algebras*, New York-London: Academic Press 1966.

G. B. SELIGMAN, [1] *Modular Lie Algebras, Ergebnisse der Mathematik und ihrer Grenzgebiete*, Band 40, Berlin-Heidelberg-New York: Springer-Verlag 1967.

[2] Topics in Lie algebras, mimeographed lecture notes, New Haven, Conn.: Yale Univ. Math. Dept. 1969.

Séminaire "Sophus Lie", [1] *Théorie des algèbres de Lie. Topologie des groupes de Lie*, Paris: École Norm. Sup. 1954–55.

J.-P. SERRE, [1] *Lie Algebras and Lie Groups*, New York: W. A. Benjamin 1965.

[2] *Algèbres de Lie semi-simples complexes*, New York: W. A. Benjamin 1966.

T. A. SPRINGER, [1] Weyl's character formula for algebraic groups, *Invent. Math.* **5**, 85–105 (1968).

R. STEINBERG, [1] A general Clebsch-Gordan theorem, *Bull. Amer. Math. Soc.* **67**, 406–407 (1961).

[2] Lectures on Chevalley groups, mimeographed lecture notes, New Haven, Conn.: Yale Univ. Math. Dept. 1968.

J. TITS, [1] Sur les constantes de structure et le théorème d'existence des algèbres de Lie semi-simples, *Inst. Hautes Études Sci. Publ. Math. No. 31*, 21–58 (1966).

V. S. VARADARAJAN, [1] Lie Groups, mimeographed lecture notes, Los Angeles: U.C.L.A. Math. Dept. 1967.

D.-N. VERMA, [1] Structure of certain induced representations of complex semi-simple Lie algebras, Yale Univ., dissertation, 1966; cf. *Bull Amer. Math Soc.* **74**, 160–166 (1968).

[2] Möbius inversion for the Bruhat ordering on a Weyl group, *Ann. Sci. École Norm. Sup. 4e série*, t.**4**, 393–398 (1971).

D. J. WINTER, [1] *Abstract Lie Algebras*, Cambridge, Mass.: M.I.T. Press 1972.

W. J. WONG, [1] Irreducible modular representations of finite Chevalley groups, *J. Algebra* **20**, 355–367 (1972).

Index of Terminology

Index of Symbols

Graduate Texts in Mathematics

Soft and hard cover editions are available for each volume.

Vol. 1 TAKEUTI/ZARING: Introduction to Axiomatic Set Theory.
vii, 250 pages. 1971.

Vol. 2 OXTOBY: Measure and Category. viii, 95 pages. 1971.

Vol. 3 SCHAEFER: Topological Vector Spaces. xi, 294 pages. 1971.

Vol. 4 HILTON/STAMMBACH: A Course in Homological Algebra.
ix, 338 pages. 1971.

Vol. 5 MAC LANE: Categories for the Working Mathematician.
ix, 262 pages. 1972.

Vol. 9 HUMPHREYS: Introduction to Lie Algebras and Representation Theory.
xiv, 169 pages. 1972.

In preparation

Vol. 6 HUGHES/PIPER: Projective Planes. xii, 296 pages approximately.
Tentative publication date: March, 1973.

Vol. 7 SERRE: A Course in Arithmetic. x, 115 pages approximately.
Tentative publication date: February, 1973.

Vol. 8 TAKEUTI/ZARING: Axiomatic Set Theory. x, 236 pages approximately.
Tentative publication date: February, 1973.

Vol. 10 COHEN: A Course in Simple-Homotopy Theory. xii, 112 pages
approximately. Tentative publication date: February, 1973.

Vol. 11 CONWAY: Complex Functions Theory. x, 313 pages approximately.
Tentative publication date: July, 1973.